FAUNA AND FAMILY

ALSO BY GERALD DURRELL

Fillets of Plaice (Godine, 2008)
Beasts in My Belfry (Godine, 2016)
The Overloaded Ark
The Bafut Beagles
Three Tickets to Adventure
The Drunken Forest
My Family and Other Animals
A Zoo in My Luggage
The Whispering Land
Menagerie Manor
Two in the Bush
Birds, Beasts and Relatives
Rosy Is My Relative (a novel)

FOR CHILDREN

The New Noah
The Donkey Rustlers

GERALD DURRELL

Fauna and Family

More Durrell Family Adventures on Corfu

A Nonpareil Book
David R. Godine · Publisher · Boston

This is a Nonpareil Book
published in 2012 by
DAVID R. GODINE, *Publisher*
Post Office Box 450
Jaffrey, New Hampshire 03452
www.godine.com

LIBRARY OF CONGRESS CATALOGING-IN-PUBLICATION DATA

Durrell, Gerald, 1925–1995.
[Garden of the gods]
Fauna and family : more Durrell Family Adventures on Corfu
/ by Gerald Durrell.
p. cm.
Originally published: The garden of the gods.
London : Collins, 1978.
ISBN-13: 978-1-56792-441-1
ISBN-10: 1-56792-441-7
1. Durrell, Gerald, 1925–1995.
2. Natural history—Greece—Corfu Island.
3. Corfu Island (Greece)—Description and travel.
4. Zoologists—Great Britain—Biography.
I. Title.
QL31.D87A33 2012
508.495'5—dc22
2010051655

THIRD PRINTING, 2016
Printed in the United States of America

Contents

This book is for Ann Peters, at one time my secretary and always my friend, because she loves Corfu and probably knows it better than I do.

A Word in Advance

THIS IS THE THIRD BOOK that I have written about a sojourn which my family and I had on the island of Corfu before the last world war. It may seem curious to some people that I can still find material to write about this period of my life; however, may I point out that we were in those days, and certainly by Greek standards, comparatively wealthy; none of us worked in the accepted sense of the word, and therefore most of our time was spent having fun. If you have five years of doing this, you accumulate quite a lot of experiences.

The pitfall of writing a series of books about the same, or essentially the same, characters, is that you do not want to bore a reader of your previous books with endless descriptions of the characters whom he knows. At the same time, you cannot be so vain as to suppose that everyone has read those previous books and so you must assume to a certain extent that the reader is approaching your work for the first time. It is difficult, therefore, to steer a course between irritating your old reader and overburdening your new one. I hope I have succeeded in doing that.

In the first book of the trilogy which I wrote – *My Family and Other Animals* – I had the following thing to say about it, which I don't think I can better: "I have attempted to draw an accurate and unexaggerated picture of my family in the following pages; they appear as I saw them. To explain some of their more curious ways, however, I feel that I should state that at the time we were in Corfu the family were all quite young: Larry, the eldest, was twenty-three; Leslie was nineteen; Margo, eighteen; while I was the youngest, being of the tender and impressionable age of ten. We have never been very certain of my mother's age, for the simple reason that she can never remember her date of birth; all I can say is that she was old enough to have four children. My mother also insists that I explain that she is a widow, for, as she so penetratingly observed, you never know what people might think.

In order to compress five years of incident, observation, and pleasant living into something a little less lengthy than the *Encyclopædia Britannica*, I have been forced to telescope, prune, and graft, so that there is little left of the continuity of events."

I also said that I had left out a number of incidents and characters that I would have liked to have described, and I have attempted to repair this omission in this book. I hope that it might give the same pleasure to its readers as apparently its predecessors – *My Family and Other Animals* and *Birds, Beasts and Relatives* – have done, as for me it portrays a very important part of my life and the thing which, unfortunately, a lot of children nowadays seem to lack, which is a truly happy and sunlit childhood.

FAUNA AND FAMILY

The Garden of the Gods

Behold! the heavens do ope,
The gods look down, and this unnatural scene
They laugh at.

SHAKESPEARE, *Coriolanus*

THE ISLAND LAY bent like a misshapen bow, its two tips nearly touching the Greek and Albanian coastlines, and the blue waters of the Ionian Sea were caught in its curve like a blue lake. Outside our villa was a wide flagstoned verandah roofed with an ancient vine from which the great green clusters of grapes hung like chandeliers; from here one looked out over the sunken garden full of tangerine trees and the silver-green olive groves to the sea, blue and smooth as a flower petal. In fine weather we always had our meals on the verandah at the rickety marble-topped table, and it was here that all the major family decisions were taken.

It was at breakfast time that there was liable to be the most acrimony and dissension, for it was then that letters, if any, were read and plans for the day were made, remade and discarded; it was during these early-morning sessions that the family fortunes were organized, albeit haphazardly, so that a simple request for an omelette might end in a three-month camping expedition to a remote beach, as had happened on one occasion.

So when we assembled in the brittle morning light, one was never quite sure how the day was going to get on its feet. To begin with, one had to step warily, for tempers were fragile, but gradually, under the influence of tea, coffee, toast, homemade marmalade, eggs and bowls of fruit, a lessening of the early-morning tension would be felt and a more benign atmosphere would begin to permeate the verandah.

The morning that heralded the arrival of the count among us was no different from any other. We had all reached the final cup of coffee stage, and each was busy with his own thoughts; Margo, my sister, her blonde hair done up in a bandana, was musing over two pattern books, humming gaily but tunelessly to herself; Leslie had finished his coffee and produced a small automatic pistol from his pocket, dismantled it, and was absentmindedly cleaning it with his handkerchief; my mother was perusing the pages of a cookery book in pursuit of a recipe for lunch, her lips moving soundlessly, occasionally breaking off to stare into space while she tried to remember if she had the necessary ingredients for the recipe she was reading; Larry, my elder brother, clad in a multicolored dressing gown, was eating cherries with one hand and reading his mail with the other.

I was occupied feeding my latest acquisition, a young jackdaw, who was such a singularly slow eater I had christened him Gladstone, having been told that that statesman always chewed everything several hundred times. While waiting for him to digest each mouthful, I stared down the hill at the beckoning sea and planned my day. Should I make a trip to the high olive groves in the center of the island to try and catch the agamas that lived on the glittering gypsum cliffs, where they basked in the sun, tantalizing me by wagging their yellow heads at me and puffing out their orange throats? Or should I go down to the small lake in the valley behind the villa, where the dragonfly larvae should be hatching? Or should I perhaps – happiest thought of all – take my new boat on a major sea trip?

In spring the almost enclosed sheet of water that separated Corfu from the mainland would be a pale and delicate blue, and then as spring settled into hot, crackling summer, it seemed to stain the still sea a deeper and more unreal color, which in some lights seemed like the violet blue of a rainbow, a blue that faded to a rich jade green in the shallows. In the evening when the sun sank, it was as if it were drawing a brush across the sea's surface, streaking and blurring it to purples smudged with gold, silver, tangerine and pale pink. To look at this placid, land-locked sea in summer when it seemed so mild-mannered, a blue meadow that breathed gently and evenly along the shoreline, it was difficult to believe that it could be fierce; but even on a still, summer's day, somewhere in the eroded hills of the mainland, hot fierce wind would suddenly be born and leap, screaming, at the island, turning the sea so dark it was almost black, combing each wave crest into a sheaf of white froth and urging and harrying them like a herd of panic-stricken blue horses until they crashed exhausted on the shore and died in a hissing shroud of foam. And in winter, under an iron-grey sky, the sea would lift sullen muscles of almost colorless waves, ice-cold and unfriendly, veined here and there with mud and debris that the winter rains swept out of the valleys and into the bay. To me, this blue kingdom was a treasure house of strange beasts which I longed to collect and observe, and at first it was frustrating for I could only peck along the shoreline like some forlorn seabird, capturing the small fry in the shallows and occasionally being tantalized by something mysterious and wonderful cast up on the shore. But then I got my boat, the good ship *Bootle Bumtrinket*, and so the whole of this kingdom was opened up for me, from the golden red castles of rock and their deep pools and underwater caves in the north to the long, glittering white sand dunes lying like snowdrifts in the south.

I decided on the sea trip, and so intent was I on planning it that I had quite forgotten Gladstone, who was wheezing at me with the breathless indignation of an asthmatic in a fog.

"If you *must* keep that harmonium covered with feathers," said Larry, glancing up irritably, "you might at least teach it to sing properly."

He was obviously not in the mood to receive a lecture on the jackdaw's singing abilities, so I kept quiet and shut Gladstone up with a mammoth mouthful of food.

"Marco's sending Count Rossignol for a couple of days," Larry said casually to Mother.

"Who's he?" asked Mother.

"I don't know," said Larry.

Mother straightened her glasses and looked at him.

"What do you mean, you don't know?" she asked.

"What I say," said Larry. "I don't know; I've never met him."

"Well, who's Marco?"

"I don't know; I've never met him either. He's a good artist though."

"Larry, dear, you can't start inviting people you don't know to stay," said Mother. "It's bad enough entertaining the ones you *do* know without starting on the ones you don't know."

"What's knowing them got to do with it?" asked Larry, puzzled.

"Well, if you know them, at least they know what to expect," Mother pointed out.

"Expect?" said Larry coldly. "You'd think I was inviting them to stay in a ghetto or something, the way you go on."

"No, no, dear, I don't mean *that*," said Mother, "but it's just that this house so seldom seems normal. I do try, but we don't seem able to live like other people somehow."

"Well, if they come to stay here, they must put up with us," said Larry. "And anyway, you can't blame *me*; I didn't invite him. Marco's sending him."

"But that's what I mean," said Mother. "Complete strangers sending complete strangers to us, as if we were a hotel or something."

"Trouble with you is you're antisocial," said Larry.

"And so would you be if you had to do the cooking," said Mother indignantly. "It's enough to make one want to be a hermit."

"Well, as soon as the count's been, you can be a hermit if you want to," said Larry. "No one's stopping you."

"A lot of chance *I* get to be a hermit, with you inviting streams of people to stay."

"Of course you can, if you organize yourself," said Larry. "Leslie will build you a cave down in the olive groves; you can get Margo to stitch a few of Gerry's less smelly animal skins together to wear, a pot of blackberries, and there you are. I can bring people down to see you. 'This is my mother,' I shall say. 'She has deserted us to become a hermit.'"

Mother glared at him.

"Really, Larry, you do make me cross sometimes," she said.

"I'm going down to see Leonora's baby," said Margo. "Is there anything you want from the village?"

"Oh, yes," said Larry, "that reminds me. Leonora's asked me to be a godparent to the brat."

Leonora was our maid Lugaretzia's daughter, who used to come up to the house and help us when we had a party and who, because of her sparkling good looks, was a great favorite of Larry's.

"You? A godfather?" said Margo in astonishment. "I thought godfathers were supposed to be pure and religious and things."

"How nice of her," said Mother doubtfully. "But it's a bit odd, isn't it?"

"Not half so odd as it would be if she asked him to be father," said Leslie.

"Leslie, dear, don't say things like that in front of Gerry, even in fun," said Mother. "Are you going to accept, Larry?"

"Yes," said Larry. "Why shouldn't the poor little thing have the benefit of my guidance?"

"Ha!" said Margo derisively. "Well, I shall tell Leonora that if she thinks you're going to be pure and religious, she's trying to make a pig's poke out of a sow's ear."

"If you can translate that into Greek, you're welcome to tell her," said Larry.

"My Greek's just as good as yours," said Margo belligerently.

"Now, now, dears, don't quarrel," said Mother. "I do wish you wouldn't clean your guns with your handkerchief, Leslie; the oil is impossible to get out."

"Well, I've got to clean them with *some*thing," said Leslie aggrievedly.

At this point I told Mother I was going to spend the day exploring the coast and could I have a picnic?

"Yes, dear," she said absently, "tell Lugaretzia to organize something for you. But do be careful, dear, and don't go into very deep water. Don't catch a chill and . . . watch out for sharks."

To Mother, every sea, no matter how shallow or benign, was an evil and tumultuous body of water, full of tidal waves, waterspouts, typhoons, whirlpools, inhabited entirely by giant octopus and squids and savage, sabre-toothed sharks, all of whom had the killing and eating of one or other of her progeny as their main objective in life. Assuring her that I would take great care, I hurried off to the kitchen and collected the food for myself and my animals, assembled my collecting equipment, whistled the dogs, and set off down the hill to the jetty where my boat was moored.

The *Bootle Bumtrinket*, being Leslie's first effort in boat building, was almost circular and flat-bottomed, so that, with her attractive color scheme of orange and white stripes, she looked not unlike an ornate celluloid duck. She was a friendly, stalwart craft, but owing to her shape and her lack of keel she became very flustered in anything like a heavy sea and would threaten to turn upside down and proceed that way, a thing she was very prone to do in moments of stress, so when I went on any long expeditions in her, I always took plenty of food and water in case we were blown off course and shipwrecked, and I hugged the coastline as much as possible so that I could make a dash for safety should the *Bootle Bumtrinket* be assaulted by a sudden sirocco. Owing to my boat's shape, she could not wear a tall mast without turning over and so her pocket-handkerchief-sized sail could only garner and harvest the tiniest cupfuls of wind; thus, for the most part, she was propelled from point to point with oars, and when we had a full crew on board – three dogs, an owl, and sometimes a pigeon – and were carrying a full cargo – some two dozen containers full of seawater and specimens – she was a back-aching load to push through the water.

Roger was a fine dog to take to sea, and he thoroughly enjoyed it; he also took a deep and intelligent interest in marine life and would lie for hours, with ears pricked, watching the strange convolutions of the brittle starfish in a collecting bottle. Widdle and Puke, on the other hand, were not sea dogs and were really most at home tracking down some not too fierce quarry in the myrtle groves; when they came to sea they tried to be helpful but rarely succeeded and in a crisis would start howling or jumping overboard, or, if thirsty, drinking seawater and then vomiting over your feet just as you were doing a tricky bit of navigation. I could never really tell if Ulysses, my scops owl, liked sea trips; he would sit dutifully wherever I placed him, his eyes half closed, wings pulled in, looking like one of the more malevolent carvings of Oriental deities. My pigeon, Quilp – he was the son of my original pigeon, Quasimodo – adored boating; he would take over the *Bootle Bumtrinket*'s minute foredeck and carry on as though it were the promenade deck of the *Queen Mary*. He would pace up and down, pausing to do a quick waltz occasionally, and with pouting chest would give a quick contralto concert, looking strangely like a large opera singer on a sea voyage. Only if the weather became inclement would he get nervous and would then fly down and nestle in the captain's lap for solace.

On this particular day I had decided to pay a visit to a small bay one side of which was formed by a tiny island surrounded by reefs in which there dwelt a host of fascinating creatures. My particular quarry was a peacock blenny, which I knew lived in profusion in that shallow water. Blennies are curious-looking fish with elongated bodies, some four inches long, shaped rather like an eel, and with their pop eyes and thick lips they are vaguely reminiscent of a hippopotamus. In the breeding season the males become most colorful, with a dark spot behind the eyes edged with sky blue, a dull orange humplike crest on the head and a darkish body covered with ultramarine or violet spots. The throat was pale sea green with darkish stripes on it. In contrast, the females were a light olive green with pale blue spots and leaf-green fins. I was anxious to capture some of these colorful little fish, since it was their breeding season and I was hoping to try and establish a colony of them in one of my aquariums so that I could watch their courtship. After half an hour's stiff rowing we reached the bay, which was rimmed with silvery olive groves and great golden tangles of broom that sent its heavy musky scent out over the still, clear waters. I anchored the *Bootle Bumtrinket* in two feet of water near the reef, and then taking off my clothes, and armed with my butterfly net and a wide-mouthed jar, I stepped into the gin-clear sea, which was as warm as a bath.

Everywhere there was such a profusion of life that it required stern concentration not to be diverted from one's task. Here the sea slugs, like huge warty brown sausages, lay in battalions among the multicolored weeds. On the rocks were the dark purple and black pincushions of the sea urchins, their spines turning to and fro like compass needles. Here and there, stuck to the rocks like enlarged wood lice, were the chitons and the brightly freckled top shells, moving about, each containing either its rightful owner or else a usurper in the shape of a red-faced, scarlet-clawed hermit crab. Here and there a small weed-covered rock would suddenly walk away from under your foot, revealing itself as a spider crab, with his back a neatly planted garden of weeds, to camouflage him from his enemies.

Soon I came to the area of the bay that I knew the blennies favored. It was not long before I spotted a fine male, brilliant and almost iridescent in his courting outfit of many colors. Cautiously I edged my net towards him, and he retreated suspiciously, gulping at me with his pouting lips. I made a sudden sweep with the net, but he was too wary and avoided it with ease. Several times I tried and failed, and after each attempt he retreated a little further. Finally, tiring of my attentions, he flipped off and took refuge in his home, which was the broken half of a terra-cotta pot of the sort that fishermen put down to trap unwary octopuses in. Although he was under the impression he had reached safety, it was in fact his undoing, for I simply scooped him up, pot and all, in my net and then transferred him and his home to one of my bigger containers in the boat.

Flushed with success, I continued my hunting, and by lunchtime I had caught two green wives for my blenny, as well as a baby cuttlefish and an interesting species of starfish, which I had not seen previously. The sun was now blistering hot, and most of the sea life had disappeared under rocks to lurk in the shade. I went on shore then, to sit under the olive trees to eat my lunch. The air was heavy with broom scent and full of the zinging cries of the cicadas.

As I ate, I watched a huge dragon-green lizard with bright blue eye markings along his body carefully stalk and catch a black-and-white-striped swallowtail butterfly. No mean feat, since the swallowtail rarely sits still for long and their flight is erratic and unpredictable. Moreover the lizard caught the butterfly on the wing, leaping some sixteen inches off the ground to do so. Presently, having finished my lunch, I loaded up the boat and, getting my canine crew on board, commenced to row home so that I could settle my blennies in their aquarium. Reaching the villa, I placed the male blenny, together with his pot, in the center of the larger of my aquariums and then carefully introduced the two females. Although I watched them for the rest of the afternoon, they did nothing spectacular. The male merely lay, gulping and pouting, in the entrance of his pot, and the females lay, gulping and pouting, at either end of the aquarium.

The following morning when I got up, I found, to my intense annoyance, that the blennies must have been active at dawn, for a number of eggs had been laid on the roof of the pot. Which female was responsible for this I did not know, but the male was a very protective and resolute father, attacking my finger ferociously when I picked up the pot to look at the eggs.

Determined not to miss any of the drama, I rushed and got my breakfast and ate it squatting in front of the aquarium, my gaze fixed on the blennies. The family, who had hitherto regarded my fish as the least of potential troublemakers among my pets, began to have doubts about the blennies, for as the morning wore on I would importune each passing member of the household to bring me an orange, or a drink of water, or to sharpen my pencil for me, for I was whiling away the time drawing the blennies in my diary. My lunch was served to me at the aquarium, and as the long, hot afternoon wore on, I began to feel sleepy. The dogs, long since bored with a vigil they could not understand, had gone off into the olive groves and left me and the blennies to our own devices.

The male blenny was deep in his pot, scarcely visible. One of the females had wedged herself behind some small rocks, while the other sat gulping on the sand. Occupying the aquarium with the fish were two small spider crabs, each encrusted with weeds, and one wearing a small, pink sea anemone like a rakish bonnet on his head. It was this crab who really precipitated the whole romance of the blennies. He was wandering about the floor of the aquarium, delicately popping bits of debris into his mouth with his claws, like a finicky spinster eating cucumber sandwiches, and he happened to wander up to the entrance of the pot. Immediately the male blenny emerged, glowing with iridescent colors, ready for battle. He swooped down onto the spider crab and bit at it viciously time after time. The crab, after a few ineffectual attempts to ward off the fish with its claws, meekly gave in and turned tail and scuttled off. This left the blenny, glowing virtuously, as the victor, and he sat just outside his pot looking rather smug. Now a very unexpected thing happened. The female on the sand had had her attention attracted by the fight with the crab and now she swam over and stopped some four or five inches away from the male. At the sight of her, he became very excited and his coloring seemed to glow all the more. Then, to my astonishment, he attacked the female. He dashed at her and bit at her head, at the same time curving his body like a bow and giving her blows with his tail. I watched this behavior in amazement until I suddenly realized that throughout this beating and buffeting the female was completely passive and made no attempt at retaliation. What I was witnessing was not an unprovoked attack, but a rather belligerent courtship display. As I watched I saw that, with slaps from his tail and bites at the female's head, the male blenny was in fact herding her towards his pot as a sheep dog herds sheep. Realizing that once they entered the pot I should lose sight of them, I dashed into the house and came back with an instrument I normally used for examining birds' nests. It was a bamboo pole with a small mirror set at an angle on the end. If there was a bird's nest out of reach, you could use the mirror on the end as a sort of periscope to enable you to examine the eggs or fledglings. Now I used it in the same way, but upside down. By the time I got back, the blennies were just disappearing into the pot. With great caution, so as not to disturb them, I lowered the mirror on the bamboo into the water and maneuvered it until it was at the entrance of the pot. When I had jiggled it into position, I found not only that I got a very good view of the interior of the pot but that the sunlight reflected off the mirror lit up the inside beautifully.

To begin with, the two fish stayed quite close together and there was a lot of fin waving but nothing much else. The male's attacks on the female, now she was safely in the pot, ceased, and he seemed more conciliatory towards her. After about ten minutes the female moved from the position alongside him and then proceeded to lay a small cluster of transparent eggs, which stuck to the smooth side of the pot like frog spawn. This done, she moved and the male took up his position over the eggs. Unfortunately, the female got between me and him, so I could not see him actually fertilize the eggs, but it was obvious that that was what he was doing. Then the female, feeling that her part of the procedure was over, swam out of the pot and across the aquarium, displaying no further interest in the eggs. The male, however, spent some time fussing around them and then came to lie in the mouth of the pot on guard.

I waited eagerly for the baby blennies to appear, but there must have been something wrong with the aeration of the water, for only two of the eggs hatched. One of the diminutive babies was, to my horror, eaten by his own mother, before my very eyes. Not wishing to have a double case of infanticide on my conscience, and lacking aquarium space, I put the baby in a jar and rowed down the coast to the bay where I had caught his parents. Here I released him with my blessing, in the clear tepid water ringed with golden broom, where I hoped that he would rear many multicolored offspring of his own.

THREE DAYS LATER the count appeared. He was tall and slender, with tightly curled hair as golden as a silkworm's cocoon, shining with oil, a delicately curled moustache of a similar hue, and slightly protuberant eyes of a very pale and unpleasant green. He alarmed Mother by arriving with a huge wardrobe trunk, and she was convinced that he had come to stay for the summer. But we soon found that the count found himself so attractive he felt it necessary to change his clothes about eight times a day to do justice to himself. His clothes were such elegant confections, beautifully hand-stitched and of such exquisite materials, that Margo was torn between envy at the count's wardrobe and disgust at his effeminacy. Combined with this narcissistic preoccupation with himself, the count had other equally objectionable characteristics. He drenched himself in a scent so thick it was almost visible and he had only to spend a second in a room to permeate the whole atmosphere, while the cushions he leaned against and the chairs he sat in reeked for days afterwards. His English was limited, but this did not prevent him from expounding on any subject with a sort of sneering dogmatism that made everyone's hackles rise. His philosophy, if any, could be summed up in the phrase "We do it better in France," which he used repeatedly about everything. He had such a thoroughly Gallic interest in the edibility of everything he came in contact with that one could have been pardoned for thinking him the reincarnation of a goat.

He arrived, unfortunately, in time for lunch, and by the end of the meal, without really trying, he had succeeded in alienating everybody including the dogs. It was in its way quite a *tour de force* to be able to irritate and insult five people of such different character with such ease and, apparently, without even being aware of doing it, inside two hours of arrival at a new locale. During the course of lunch, he said, having just eaten a soufflé as delicate as a cloud in which were embedded the pale pink bodies of freshly caught shrimps, that it was quite obvious that Mother's chef was not French. Having discovered that Mother was the chef, he showed no embarrassment but merely said that she would then be glad of his presence for it would enable him to give her some guidance in the culinary arts. Leaving her speechless with rage at his audacity, he turned his attention to Larry, to whom he vouchsafed the information that the only good writers were French. At the mention of Shakespeare, he merely shrugged; "*le petit poseur,*" he said. To Leslie he offered the information that anyone who was interested in hunting must assuredly have the instincts of a criminal and, in any case, it was well known that the French produced the best guns, swords and other weapons of offense. To Margo he gave the advice that it was a woman's job to keep beautiful for men and, in particular, not to be greedy and eat too many things that would ruin the figure. As Margo was suffering from a certain amount of puppy fat at that time and was on a rigid diet in consequence, this information was not at all well received. He merely condemned himself in my eyes by calling the dogs village curs and comparing them unfavorably to his selection of Labradors, setters, retrievers and spaniels, all French-bred, of course. Furthermore, he was puzzled as to why I kept so many pets, all of which were uneatable. "In French we only shoot zis kind of thing," he said.

Small wonder, then, that after lunch when he went upstairs to change, the family were quivering like a suppressed volcano, and only Mother's golden rule that a guest must not be insulted on the first day kept us in check. But such was the state of our nerves that if anyone had started to whistle the "Marseillaise," we would have torn him limb from limb.

"You see," said Mother accusingly to Larry, "this is what comes of letting people you don't know send people you don't know to stay. The man's insufferable!"

"Well . . . he's not so bad," said Larry feebly, trying to argue against an attitude that he agreed with. "I thought some of his comments were valid."

"What?" asked Mother ominously.

"Yes, what?" asked Margo, quivering.

"Well," said Larry vaguely, "I thought that soufflé was a bit on the rich side, and Margo is beginning to look a bit circular."

"Beast!" said Margo, and burst into tears.

"Now that's quite enough, Larry," said Mother. "How we're going to endure this . . . this . . . scented *lounge lizard* of yours for another week I don't know."

"Well, I've got to put up with him too, don't forget," said Larry, irritated.

"Well, he's *your* friend . . . I mean, your friend's friend . . . I mean, well, whatever he is, he's *yours*," said Mother, "and it's up to you to keep him out of the way as much as possible."

"Or I'll pepper his arse for him," said Leslie, "the smelly little – "

"Leslie," said Mother, "that's quite enough."

"Well, he is," said Leslie doggedly.

"I know he is, dear, but you shouldn't say so," Mother explained.

"Well, I'll try," said Larry, "but don't blame me if he comes down to the kitchen to give you a cookery lesson."

"I'm warning you," said Mother mutinously, "if that man sets foot in my kitchen, I shall walk out . . . I shall go . . . I shall go and. . . ."

"Be a hermit?" suggested Larry.

"No, I shall go and stay in an hotel until he's gone," said Mother, uttering her favorite threat, "and this time I really *mean* it."

To give Larry his due, he did strive manfully with Count Rossignol for the next few days. He took him to the library and museum in town, he showed him the kaiser's summer palace with all its repulsive statuary, he even took him to the top of the highest point in Corfu, Mount Pantocrator, and showed him the view. The count compared the library unfavorably with the Bibliothèque Nationale, said that the museum was not a patch on the Louvre, said the kaiser's palace was inferior in size, design and furnishings to the cottage he had for his head gardener, and said that the view from Pantocrator was not to be mentioned in the same breath with *any* view to be seen from *any* high spot in France.

"The man's intolerable," said Larry, refreshing himself with brandy in Mother's bedroom, where we had all repaired to escape the count's company. "He's got an obsession with France; I can't think why he ever left the place. He even thinks their telephone service is the best in the world! And he's so *humorless* about everything, one would think he were a Swede."

"Never mind, dear," said Mother, "it's not for long now."

"I'm not sure I shall last the course," said Larry. "So far about the only thing he hasn't claimed for France is God."

"Ah, but they probably believe in Him better in France," Leslie pointed out.

"Wouldn't it be wonderful if we could do something really nasty to him?" said Margo wistfully. "Something really horrible."

"No, Margo," said Mother firmly, "we've never done anything nasty to anyone that's stayed with us – I mean, except as a joke or by accident – and we're not going to start. We'll just have to put up with him; after all, it's only for a few more days. It'll soon pass."

"Dear God!" said Larry suddenly. "I've just remembered. It's the bloody christening on Monday!"

"I do wish you wouldn't swear so much," said Mother. "What's that got to do with it?"

"Can you imagine taking him to a *christening?*" asked Larry. "No, he'll just have to go off somewhere on his own."

"I don't think we ought to let him go wandering off on his own," said Mother, as if she were talking about a dangerous animal. "I mean, he might meet one of our friends."

We all sat and thought about the problem.

"Why doesn't Gerry take him somewhere?" said Leslie suddenly. "After all, he doesn't want to come to a boring christening."

"That's a brain wave," said Mother delightedly. "The very thing!"

Immediately all my instincts for self-preservation came to the fore. I said that I *did* want to go to the christening, I had been looking forward to it, it was the only chance I would ever have of seeing Larry being a godfather, and he might drop the baby or something and I would miss it; and anyway, the count did not like snakes and tortoises and birds and things, so what could *I* do with him? There was silence while the family, like a jury, examined the strength of my case.

"I know, take him out in your boat," suggested Margo brightly.

"Excellent!" said Larry, "I'm sure he's got a straw hat and a striped blazer among his sartorial effects. Perhaps we can borrow a banjo."

"It's a very good idea,' said Mother. "After all, it's only for a couple of hours, dear. You surely wouldn't mind that."

I stated in no uncertain terms that I would mind it very much indeed.

"I tell you what," said Leslie, "they're having a fish drive down at the lake on Monday. If I get the chap who's in charge to let you go, will you take the count?"

I wavered, for I had long wanted to see a fish drive. I knew I was going to have the count for the afternoon, so it was simply a matter of what I could get out of it.

"And then we can see about that new butterfly cabinet you want," said Mother.

"And Margo and I will give you some money for books," said Larry, generously anticipating Margo's participation in the bribery.

"And I'll give you that clasp knife you wanted," said Leslie.

I agreed. I felt that if I had to put up with the count for an afternoon, I was at least being fairly compensated for it. That evening at dinner, Mother explained the situation and went into such detailed eulogies about the fish drives that you would have thought she had personally invented them.

"Ees eating?" asked the count.

"Yes, yes," said Mother. "The fish are called *kefalias* and they're delicious."

"No, ees eating on ze lack?" asked the count. "Ees eating wiz sun?"

"Oh . . . oh, I see," said Mother. "Yes, it's very hot. Be sure to wear a hat."

"We go in ze *enfant*'s yacht?" asked the count, who liked to get things clear.

"Yes," said Mother.

The count outfitted himself for the expedition in pale blue linen trousers, elegant chestnut-bright shoes, a white silk shirt with a blue and gold cravat knotted carelessly at the throat, and an elegant yachting cap. While the *Bootle Bumtrinket* was ideal for my purposes, I would have been the first to admit that she had none of the refinements of an oceangoing yacht, and this the count was quick to perceive when I led him down to the canal in the maze of old Venetian salt pans below the house, where I had the boat moored.

"Zis . . . is yacht?" he asked in surprise and some alarm.

I said that indeed this was our craft, stalwart and stable, and, he would note, a flat bottom to make it easier to walk about in. Whether he understood me, I do not know; perhaps he thought the *Bootle Bumtrinket* was merely the dinghy in which he was to be rowed out to the yacht, but he climbed in delicately, spread his handkerchief fastidiously over the seat and sat down gingerly. I leapt aboard and with the aid of a pole started punting the craft down the canal, which at this point was some twenty feet wide and two feet deep. I congratulated myself on the fact that only the day before I had decided that the *Bootle Bumtrinket* was starting to smell almost as pungently as the count, for over a period a lot of dead shrimps, seaweed and other debris had collected under her boards. I had sunk her in some two feet of sea water and given her bilges a thorough cleaning, so now, for this expedition, she was sparkling clean and smelt beautifully of sun-hot tar and paint and salt water.

The old salt pans lay along the edge of the brackish lake, forming a giant chessboard with the cross-hatching of these placid canals, some as narrow as a chair, some thirty feet wide. Most of these waterways were only a couple of feet deep, but below the water lay an almost unplumbable depth of fine black silt. The *Bootle Bumtrinket*, by virtue of her shape and flat bottom, could be propelled up and down these inland waterways with comparative ease, for one did not have to worry about sudden gusts of wind or a sudden, bouncing cluster of wavelets, two things that always made her a bit alarmed. But the disadvantage of the canals was that they were fringed on each side with tall, rustling bamboo breaks which, while providing shade, precluded the wind, so the atmosphere was still, dark, hot and as richly odoriferous as a manure heap. For a time the artificial smell of the count vied with the scents of nature, but eventually nature won.

"Ees smell," the count pointed out. "In France ze water ees hygiene."

I said it would not be long before we left the canal and were out on the lake, where there would be no smell.

"Ees heating," was the count's next discovery, mopping his face and moustache with a scent-drenched handkerchief, "ees heating much."

His pale face had, as a matter of fact, turned a light shade of heliotrope. I was just about to say that that problem, too, would be overcome once we reached the open lake when, to my alarm, I noticed something wrong with the *Bootle Bumtrinket*. She had settled sluggishly in the brown water and hardly moved to my punting. For a moment I could not imagine what was wrong with her; we had not run aground and I knew that there were no sandbanks in this canal. Then suddenly I noticed the swirl of water coiling up over the boards in the bottom of the boat. Surely, I thought, she could not have sprung a leak. Fascinated, I watched the water rise to engulf the bottom of the oblivious count's shoes, and I suddenly realized what must have happened. When I had cleaned out the bilges I had, of course, removed the bung in the *Bumtrinket*'s bottom to let the fresh seawater in; apparently I had not replaced it with enough care, and now the canal water was pouring into the bilges. My first thought was to pull up the boards, find the bung and replace it, but the count was now sitting with his feet in about two inches of water and it seemed to me imperative that I turn the *Bootle Bumtrinket* towards the bank while I could still maneuver a trifle and get my exquisite passenger on shore. I did not mind being deposited in the canal by the *Bootle Bumtrinket*; after all, it was not her fault and, anyway, I was always in and out of the canals like a water rat in pursuit of water snakes, terrapins, frogs and other small fry, but I knew that the count would look askance at gamboling in two feet of water and an undetermined amount of mud, so my efforts to turn the leaden, waterlogged boat towards the bank were superhuman. Gradually, I felt the dead weight of the boat responding and her bows turning sluggishly towards the shore. Inch by inch I eased her towards the bamboos, and we were within ten feet of the bank when the count noticed what was happening.

"*Mon dieu!*" he cried shrilly, "ve are submerge. My shoe is submerge. Ze boat, she ave sonk."

I stopped poling briefly to soothe the count. I told him that there was no danger; all he had to do was to sit still until I got him to the bank.

"My shoe! *Regardez* my shoe!" he said, pointing at his now dripping and discolored footwear with such an expression of outrage and horror that it was all I could do not to giggle.

A moment, I said to him, and I should have him on dry land, and indeed if he had done what I had said, this would have been the case, for I had managed to get the *Bootle Bumtrinket* to within six feet of the bamboos. But the count was worried about the state of his shoes, and this prompted him to do something very silly. He looked over his shoulder and saw land looming close, and in spite of my warning shout, he got to his feet and leapt onto the *Bootle Bumtrinket*'s minute foredeck. His intention was to leap from there to safety when I had maneuvered the boat a little closer, but he had not reckoned with the *Bootle Bumtrinket*'s temperament. A placid boat, she had nevertheless a few quirks, and one thing she did not like was anyone standing on her foredeck; she simply gave an odd sort of bucking twist, rather like a trained horse in a cowboy film, and, as it were, slid you over her shoulder. She did this to the count now.

He fell into the water with a yell, spread-eagled like an ungainly frog, and his proud yachting cap floated towards the bamboo roots while he thrashed about in a porridge of water and mud. I was filled with a mixture of alarm and delight; I was delighted that the count had fallen in – though I knew my family would never believe that I had not engineered it – but I was alarmed at the way he was thrashing about. It is an instinctive action, when finding you are in shallow water, to try to stand up, but this action only makes you sink deeper into the glutinous mud. Once, Larry had fallen into one of these canals while out shooting and had made such a fuss and got himself so deeply embedded that it had required the united efforts of Margo, Leslie and myself to get him out. If the count got himself too wedged in the canal bottom, I would not have the strength to extricate him single-handed, and by the time I got help he might well have disappeared altogether beneath the gleaming mud. I abandoned ship and leapt into the canal to help him. Firstly, I knew how to walk in mud, and secondly, I weighed only a quarter of what the count weighed so I did not sink in so far. I shouted to him to keep quite still until I got to him.

"*Merde!*" said the count, proving that he was at least keeping his mouth above water.

He tried to get up once, but at the terrible, gobbling clutch of the mud, he uttered a despairing cry like a bereaved sea gull and lay still. Indeed, he was so frightened of the mud that when I reached him and tried to pull him shorewards he screamed and shouted and accused me of trying to push him in deeper. He was so absurdly childlike that I had a fit of the giggles, and this of course only made him worse. He had relapsed into French, which he was speaking with the rapidity of a machine gun, so with my tenuous command of the language I was unable to understand him. Eventually I got my unmannerly laughter under control and once more seized him under the armpits and started to drag him shorewards, but it suddenly occurred to me how ludicrous our predicament would seem to an onlooker – a twelve-year-old boy trying to rescue a six-foot man – and I was overcome again and sat down in the mud and laughed till I cried.

"Vy you laughing? Vy you laughing?" screamed the count, trying to look over his shoulder at me. "You no laughing, you pulling, *vite, vite!*"

Eventually, swallowing great hiccups of laughter, I started to pull the count again and eventually got him fairly close to the shore. Then I left him and climbed out onto the bank. This provoked another bout of hysteria.

"No going avay! No going avay!" he yelled, panic-stricken. "I am sonk. No going avay!"

I ignored him, and choosing seven of the tallest bamboos in the vicinity, I bent them over one by one until their stems splintered but did not snap, and then I twisted them round until they reached the count and formed a sort of green bridge between him and the shore. Acting on my instructions, he turned on his stomach and pulled himself along on this until at last he reached dry land. When he eventually got somewhat shakily to his feet, he looked as though the lower half of his body had been encased in melting chocolate. Knowing that this glutinous mud could dry hard in record time, I offered to scrape some of it off him with a piece of bamboo. He gave me a murderous look.

"*Espèce de con!*" he said vehemently.

My shaky knowledge of the count's language did not allow me to translate this, but the enthusiasm with which it was uttered led me to suppose that it was worth retaining in my memory, which I did. We started to walk home, the count simmering vitriolically. As I had anticipated, the mud on his legs dried at an almost magical speed and within a short time he looked as though he were wearing a pair of trousers made out of a pale brown jigsaw puzzle. From the back, he reminded me so much of the armor-clad rear of an Indian rhinoceros that I almost got the giggles again.

It was unfortunate, perhaps, that the count and I should have arrived at the front door of the villa just as the huge Dodge driven by our scowling, barrel-shaped, self-appointed guardian angel, Spiro Halikiopoulos, drew up with the family, flushed with wine, in the back of it. The car came to a halt and the family stared at the count with disbelieving eyes. It was Spiro who recovered first.

"Gollys, Mrs. Durrells," he said, twisting his massive head round and beaming at Mother, "Master Gerrys fixes the bastards."

This was obviously the sentiment of the whole family, but Mother threw herself into the breach.

"My goodness, Count," she said in well-simulated tones of horror, "what *have* you been doing with my son?"

The count was so overcome with the audacity of this remark that he could only look at Mother open-mouthed.

"Gerry dear," Mother went on, "go and change out of those wet things before you catch cold, there's a good boy."

"Good boy!" repeated the count, shrilly and unbelievingly. "*C'est un assassin! C'est une espèce de –* "

"Now, now, my dear fellow," said Larry, throwing his arm round the count's muddy shoulders, "I'm sure it's been a mistake. Come and have a brandy and change your things. Yes, yes, rest assured that my brother will smart for this. Of course he will be punished."

Larry led the vociferous count into the house, and the rest of the family converged on me.

"What did you do to him?" asked Mother.

I said I had not done anything; the count and the count alone was responsible for his condition.

"I don't believe you," said Margo. "You always say that."

I protested that had I been responsible I would be proud to confess. The family were impressed by the logic of this.

"Well, it doesn't matter a damn if Gerry did it or not," said Leslie. "It's the end result that counts."

"Well, go and get changed, dear," said Mother, "and then come to my room and tell us all about how you did it."

But the affair of the *Bootle Bumtrinket* did not have the effect that everyone hoped for; the count stayed on grimly, as if to punish us all, and was twice as offensive as before. However, I had ceased feeling vindictive towards him; whenever I thought of him thrashing about in the canal, I was overcome with helpless laughter, which was worth any amount of insults. And, furthermore, the count had unwittingly added a fine new phrase to my French vocabulary. I tried it out one day when I made a mistake in my French composition and I found it tripped well off the tongue. The effect on my tutor, Mr. Kralefsky, was, however, very different. He had been pacing up and down the room, hands behind him, looking like a humpbacked gnome in a trance. At my expression, he came to a sudden stop, wide-eyed, looking like a gnome who had just had an electric shock from a toadstool.

"*What* did you say?" he asked in a hushed voice.

I repeated the offending phrase. Mr. Kralefsky closed his eyes, his nostrils quivered, and he shuddered.

"*Where* did you hear that?" he asked.

I said I had learned it from a count who was staying with us.

"Oh. Well, you must never say it again, do you understand," Mr. Kralefsky said, "never again! You . . . you must learn that in this life sometimes even *aristocrats* let slip an unfortunate phrase in moments of stress. It does not behoove us to imitate them."

I did see what Kralefsky meant. Falling into a canal, for a count, could be called a moment of stress, I supposed.

But the saga of the count was not yet over. A week or so after he had departed, Larry, one morning at breakfast, confessed to feeling unwell. Mother put on her glasses and stared at him critically.

"How do you mean, unwell?" she asked.

"Not my normal, manly, vigorous self," said Larry.

"Have you got any pains?" asked Mother.

"No," Larry admitted, "no actual pains. Just a sort of lassitude, a feeling of ennui, a debilitated, drained feeling, as if I had spent the night with Count Dracula; and I feel that, for all his faults, our late guest was not a vampire."

"Well, you look all right," said Mother, "though we'd better get you looked at. Dr. Androccelli is on holiday, so I'll have to get Spiro to bring Theodore."

"All right," said Larry listlessly, "and you'd better tell Spiro to nip in and alert the British cemetery."

"Larry, don't say things like that," said Mother, getting alarmed. "Now, you go up to bed and, for heaven's sake, stop there."

If Spiro could be classified as our guardian angel to whom no request was impossible of fulfillment, Dr. Theodore Stephanides was our oracle and guide to all things. He arrived, sitting sedately in the back of Spiro's Dodge, immaculately dressed in a tweed suit, his homburg at just the correct angle, his beard twinkling in the sun.

"Yes, it was really . . . um . . . very curious," said Theodore, having greeted us all, "I was just thinking to myself how nice a trip . . . that is to say, a *spin* in the country would be as it . . . er . . . was an especially beautiful day . . . um . . . not too hot, and that sort of thing, you know . . . er . . . and suddenly Spiro turned up at the laboratory. Most fortuitous."

"I'm so glad that my agony is of some benefit to *someone*," said Larry.

"Aha! What . . . er . . . you know . . . seems to be the trouble?" asked Theodore, eyeing Larry with interest.

"Nothing concrete," Larry admitted, "just a general feeling of death being very imminent. All my strength seems to have drained away. I've probably, as usual, been giving too much of myself to my family."

"I don't think *that's* what's wrong with you," said Mother decisively.

"I think you've been eating too much," said Margo. "What you want is a good diet."

"What he wants is a little fresh air and exercise," said Leslie. "If he took the boat out a bit . . . "

"Yes, well, Theodore will tell us what's wrong," said Mother.

Theodore examined Larry and reappeared in half an hour's time.

"I can't find anything . . . er . . . you know . . . organically wrong," said Theodore judiciously, rising and falling on his tiptoes, "except that he is perhaps a trifle overweight."

"There you are! I told you he needed a diet," said Margo triumphantly.

"Hush, dear," said Mother. "So what do you advise, Theodore?"

"I should keep him in bed for a day or so," said Theodore. "Give him a light diet, you know, nothing very oily, and I'll send out some medicine . . . er . . . that is to say . . . a tonic for him. I'll come out the day after tomorrow and see how he is."

Spiro drove Theodore back to town and in due course reappeared with the medicine.

"I won't drink it," said Larry, eyeing the bottle askance, "it looks like essence of bat's ovaries."

"Don't be silly, dear," said Mother, pouring some into a spoon, "it will do you good."

"It won't," said Larry. "It's the same stuff that my friend Dr. Jekyll took, and look what happened to him."

"What happened to him?" asked Mother, unthinkingly.

"They found him hanging from the chandelier, scratching himself and saying he was Mr. Hyde."

"Come on now, Larry, stop fooling about," said Mother firmly.

With much fussing, Larry was prevailed upon to take the medicine and retire to bed.

The following morning we were all woken at an inordinately early hour by roars of rage coming from Larry's room.

"Mother! Mother!" he was roaring. "Come and look what you've done!"

We found him prancing around his room, naked, a large mirror in one hand. He turned on Mother belligerently, and she gasped at the sight of him. His face was swollen up to about twice normal size and was the approximate color of a tomato.

"*What* have you been doing, dear?" asked Mother faintly.

"Doing? It's what *you've* done," he shouted, articulating with difficulty. "You and bloody Theodore and your damned medicine. It's affected my pituitary. Look at me! It's worse than Jekyll and Hyde."

Mother put on her spectacles and gazed at Larry.

"It looks to me as though you've got mumps," she said, puzzled.

"Nonsense! That's a child's disease," said Larry impatiently. "No, it's that damned medicine of Theodore's. I tell you, it's affected my pituitary. If you don't get the antidote straight away, I shall grow into a giant."

"Nonsense, dear, I'm sure it's mumps," said Mother, "but it's very funny, because I'm sure you've had mumps. Let's see, Margo had measles in Darjeeling in 1920 . . . Leslie had sprue in Rangoon – no, I'm wrong, that was 1900 in Rangoon and *you* had sprue. Then Leslie had chicken pox in Bombay in 1911 . . . or was it 12? I can't quite remember. And then you had your tonsils out in Rajaputana in 1922, or it may have been 1923, I can't remember exactly, and then after that, Margo got – "

"I hate to interrupt this *Old Moore's Almanac of Family Ailments*," said Larry coldly, "but would somebody like to send for the antidote before I get so big I can't leave the room?"

Theodore, when he appeared, agreed with Mother's diagnosis.

"Yes . . . er . . . um . . . clearly a case of mumps," he said.

"What do you mean, *clearly*, you charlatan?" said Larry, glaring at him from watering and swollen eyes. "Why didn't you know what it was yesterday? And anyhow, I can't get mumps, it's a child's disease."

"No, no," said Theodore. "Children generally get it, but quite often adults get it too."

"Why didn't you recognize a common disease like that when you saw it?" demanded Larry. "Can't even recognize a mump when you see it? You ought to be drummed out of the medical council or whatever it is that they do for malpractice."

"Mumps are very difficult to diagnose in the . . . er . . . early stages," said Theodore, "until the swellings appear."

"Typical of the medical profession," said Larry bitterly.

"They can't even spot a disease until the patient is twice life-size. It's a scandal."

"As long as it doesn't affect your . . . um . . . you know . . . um . . . your . . . er . . . lower quarters," said Theodore thoughtfully, "you should be all right in a few days."

"Lower quarters?" Larry asked, mystified. "What lower quarters?"

"Well, er . . . you know . . . mumps causes swelling of the glands," explained Theodore, "and so if it travels down the body and affects the glands in your . . . um . . . lower quarters, it can be very painful indeed."

"You mean I'll swell up and start looking like a bull elephant?" asked Larry in horror.

"Mmm, er . . . yes," said Theodore, finding he could not better this description.

"It's a plot to make me sterile!" shouted Larry. "You and your bloody tincture of bat's blood! You're jealous of my virility."

To say that Larry was a bad patient would be putting it mildly. He had an enormous hand-bell by the bed which he rang incessantly for attention, and Mother had to examine his nether regions about twenty times a day to assure him that he was not in any way affected. When it was discovered that it was Leonora's baby that had given him mumps, he threatened to excommunicate it.

"I'm its godfather," he said. "Why can't I excommunicate the ungrateful little bastard?"

By the fourth day we were all beginning to feel the strain, and then Captain Creech appeared to see Larry. Captain Creech, a retired mariner of lecherous habits,was mother's *bête noire*. His determined pursuit of anything female, and Mother in particular, in spite of his seventy-odd years, was a constant source of annoyance to her, as were the captain's completely uninhibited behavior and one-track mind.

"Ahoy!" he shouted, staggering into the bedroom, his lopsided jaw waggling, his wispy beard and hair standing on end, his rheumy eyes watering. "Ahoy, there! Bring out your dead!"

Mother, who was just examining Larry for the fourth time that day, straightened up and glared at him.

"Do you mind, Captain?" she said coldly. "This is supposed to be a sickroom, not a bar parlor."

"Got you in the bedroom at last!" said Creech, beaming, taking no notice of Mother's expression. "Now, if the boy moves over; we can have a little cuddle."

"I'm far too busy to cuddle, thank you," said Mother frostily.

"Well, well," said the captain, seating himself on the bed, "what's this namby-pamby mumps thing you've got, huh, boy? Child stuff! If you want to be ill, be ill properly, like a man. Why, when I was your age, nothing but a dose of clap would have done for me."

"Captain, I would be glad if you would not reminisce in front of Gerry," said Mother firmly.

"It hasn't affected the old manhood, has it?" asked the captain with concern. "Terrible when it gets you in the crutch. Can ruin a man's sex life, mumps in the crutch."

"Larry is perfectly all right, thank you," said Mother with dignity.

"Talking of crutches," said the captain, "have you heard about the young Hindu virgin from Kutch? Who kept two tame snakes in her crutch? She said when they wriggle, it's a bit of a giggle, but my boyfriends don't like my crutch much. Ha ha ha!"

"Really, Captain!" said Mother, outraged. "I do wish you wouldn't recite poetry in front of Gerry."

"Got your mail. I was passing the post office," the captain went on, oblivious of Mother's strictures, pulling it out of his pocket and tossing it onto the bed. "My, they've got a nice little bit serving in there now. She'd win a prize for the best marrows in any horticultural show."

But Larry was not listening; he had extracted a postcard from the mail Captain Creech had brought and, having read it, he started to laugh uproariously.

"What is it, dear?" asked Mother.

"A postcard from the count," said Larry, wiping his eyes.

"Oh, *him*," sniffed Mother, "well, I don't want to know about *him*."

"Oh, yes, you will," said Larry. "It's worth being ill just to be able to get this. I'm starting to feel better already."

He picked up the postcard and read it out to us. The count had obviously got someone to write the card for him, and the person's command of English was fragile but inventive.

"I have reeching Rome," it began. "I am in clinic inflicted by disease called moops. Have inflicted all over. I finding I cannot arrange myself. I have no hunger and impossible I am sitting. Beware yourself the moops. Count Rossignol."

"Poor man," said Mother without conviction when we had all stopped laughing, "we shouldn't really laugh."

"No," said Larry. "I'm going to write and ask him if Greek moops are inferior in virulence to French moops."

The Elements of Spring

An habitation of dragons, and a court for owls.
ISAIAH 34:13

SPRING, IN ITS SEASON, came like a fever; it was as though the island shifted and turned uneasily in the warm, wet bed of winter and then, suddenly and vibrantly, was fully awake, stirring with life under a sky as blue as a hyacinth bud into which a sun would rise, wrapped in mist as fragile and as delicately yellow as a silkworm's newly completed cocoon. For me, spring was one of the best times, for all the animal life of the island was astir and the air was full of hope. Maybe today I would catch the biggest terrapin I had ever seen or fathom the mystery of how a baby tortoise, emerging from its egg as crushed and wrinkled as a walnut, would, within an hour, have swelled to twice its size and have smoothed out most of its wrinkles in consequence. The whole island was a-bustle and ringing with sound. I would awake early, breakfast hurriedly under the tangerine trees already fragrant with the warmth of the early sun, gather my nets and collecting boxes, whistle for Roger, Widdle and Puke, and set off to explore my kingdom.

Up in the hills, in the miniature forests of heather and broom, where the sun-warmed rocks were embossed with strange lichens like ancient seals, the tortoises would be emerging from their winter sleep, pushing aside the earth that they had slept under and jerking slowly out into the sun, blinking and gulping. They would rest until the sun had warmed them, and then they would move slowly off towards the first meal of clover or dandelion, or maybe a fat, white puffball. Like other parts of my territory, I had the tortoise hills well organized; each tortoise possessed a number of distinguishing marks, so I could follow its progress. Each nest of stone chats or blackcaps was carefully marked so that I could watch progress, as was each papery mound of mantis eggs, each spider's web and each rock under which lurked some beast dear to me.

But it was the heavy emergence of the tortoises that would really tell me that spring had started, for it was not until winter was truly over that they lumbered forth in search of mates, cumbersome and heavily armored as any medieval knight in search of a damsel to succor. Having once satisfied their hunger, they became more alert – if such a word can be used to describe a tortoise. The males walked on their toes, their necks stretched out to the fullest extent, and at intervals they would pause and utter an astonishing, loud and imperative yap. I never heard a female answer this ringing, Pekinese-like cry, but by some means the male would track her down and then, still yapping, do battle with her, crashing his shell against hers, trying to bludgeon her into submission, while she, undeterred, would try to go on feeding in between the bouts of buffeting. So the hills would resound to the yaps and slithering crashes of the mating tortoises and the stonechats' steady *tak tak* like a miniature quarry at work, the cries of pink-breasted chaffinches like tiny, rhythmic drops of water falling into a pool, and the gay, wheezing song of the goldfinches as they tumbled through the yellow broom like multicolored clowns.

Down below the tortoise hills, below the old olive groves filled with wine-red anemones, asphodels and pink cyclamen, where the magpies made their nests and where the jays would startle you with their sudden harsh, despairing scream, lay the old Venetian salt pans, spread out like a chessboard. Each field, some only the size of a small room, was bounded by wide, shallow, muddy canals of brackish water. Each field was a little jungle of vines, maize, fig trees, tomatoes as acrid-smelling as stinkbugs, watermelons like the huge green eggs of some mythical bird, trees of cherry, plum, apricot, and loquat, strawberry plants and sweet potatoes, all forming the larder of the island. On the seaward side, each brackish canal was fringed with canebrakes and reed beds sharply pointed as an army of pikes; but inland, where the streams fell from the olive groves into the canals and the water was sweet, you got lush plant growth and the placid canals were emblazoned with water lilies and fringed with golden kingcups.

It was here that in the spring the two species of terrapin – one black with gold spots and one pin-striped delicately with grey – would whistle shrilly, almost like birds, as they pursued their mates. The frogs, green and brown, with leopard-patched thighs, looked as though they were freshly varnished. They would clasp each other with passionate, pop-eyed fervor or gurk an endless chorus and lay great cumulous clouds of grey spawn in the water. In places where the canals were bordered by shade-giving canebrakes and fig and other fruit trees, the diminutive tree frogs, vivid green, with skin as soft as a damp chamois leather, would puff up their little yellow throat pouches to the size of walnuts and croak in a monotonous tenor voice. In the water, where the pigtails of weed moved and undulated gently in the baby currents, the tree frogs' spawn would be laid in yellowish lumps the size of a small plum.

Along one side of the fields lay a flat grassland area which, with the spring rains, would be inundated and turn into a large shallow lake some four inches deep, lined with grass. Here, in this warm water, the newts would assemble, hazelnut-brown with yellow bellies. A male would take up his station facing the female, tail curved round, and then, with a look of almost laughable concentration on his face, he would wag his tail ferociously, ejaculating sperm and wafting it towards the female. She, in her turn, would place each fertilized egg, white and almost as transparent as the water, yolk black and shining as an ant, onto a leaf and then, with her hind legs, bend the leaf and stick it so that the egg was encased.

In spring the herds of strange cattle would appear to graze on this drowned lake. Huge, chocolate-colored animals with massive, backward-slanting horns as white as mushrooms, they looked like the Ankole cattle from the center of Africa, but they must have been brought from somewhere nearer, Persia or Egypt perhaps. They were tended by strange, wild, Gypsy-like bands, who in long, low, horse-drawn wagons would camp by the grazing area; the savage-looking men, dusky as crows, and the handsome women and girls with velvety black eyes and hair like moleskin would sit gossiping or making baskets around the fire, speaking a language I could not understand, while the raggedly dressed boys, thin and brown, jay-shrill and jackal-suspicious, would act as herdsmen. The great beasts' horns would clack and rattle together like musketry as they barged each other out of the way in their eagerness to feed. The sweet cattle smell of their brown coats lingered in the warm air after them like the scent of flowers. One day the grazing area would be empty; the next day, as if they had always been there, there would be the jumbled encampment caught in a perpetual spider's web of smoke from its pink, glittering fires and the herds of cattle moving slowly through the shallow water, their probing, tearing mouths and splashing hoofs frightening the newts and sending the frogs and baby terrapins off in panic-stricken flight at this mammoth invasion.

I greatly coveted one of these huge, brown cattle, but I knew that my family would not under any circumstances allow me to have anything so large and so fierce-looking, no matter how much I pleaded that they were so tame that they were herded by boys of six or seven, mere toddlers. The nearest I got to possessing one of these animals was quite close enough as far as the family was concerned. I had been down in the fields just after the Gypsies had killed a bull; the still bloody hide was stretched out and a group of girls were scraping it with knives and rubbing wood ash into it. Nearby was piled its gory, dismembered carcass, already shining and humming with flies, and next to it was the massive head, the fringed ears lying back, the eyes half closed as if musing, a trickle of blood coming from one nostril. The sweeping white horns were some four feet long and as thick as my thigh, and I gazed at them longingly, as covetous as any early big-game hunter.

It would be impractical, I decided, to buy the whole head; even though I was convinced of my mastery of the art of taxidermy, the family did not share my conviction. Besides, there had recently been a bit of unpleasantness over a dead turtle I had unthinkingly dissected on the verandah, so everyone was inclined to view my interest in anatomy with a jaundiced eye. It was a pity, really, for the bull's head, carefully mounted, would have looked magnificent over the door of my bedroom and would have been the *pièce de résistance* of my entire collection, surpassing even my stuffed flying fish and my almost complete goat skeleton. However, knowing how implacable my family could be, I decided reluctantly I would have to settle for the horns. After a spirited piece of bargaining – the Gypsies knew enough Greek for that – I purchased the horns for ten drachmas and my shirt. The absence of the shirt I explained to Mother by saying I had ripped it so badly falling out of a tree that the remnants were not worth bringing back. Then, triumphantly, I carried the massive horns up to my room and spent the morning polishing them, nailing them to a plaque of wood, and then hanging the whole thing with great care on a hook over my door.

I stood back to admire the effect and at that moment heard Leslie's voice raised in anger.

"Gerry!" he shouted. "Gerry! Where are you?"

I remembered that I had borrowed his tin of gun oil from his room to polish the horns with, meaning to restore it before he noticed. But before I could do anything, the door burst open and he appeared belligerently.

"Gerry! Have you got my bloody gun oil?" he inquired.

The door, returning on the impetus of his entrance, swung back and slammed to. My magnificent pair of horns leapt off the wall as if propelled by the ghost of the bull that had possessed them and landed on top of Leslie's head, felling him as though he had been poleaxed.

My first thought was that my beautiful horns might be broken; my second, that my brother might be dead. Both proved to be erroneous. My horns were intact and my brother, his eyes still glazed, hoisted himself into a sitting position and stared at me.

"Christ! My head!" he moaned, clasping his temples and rocking to and fro. "Bloody hell!"

As much to dilute his wrath as anything, I went in search of Mother. I found her in her bedroom brooding over the bed, which was covered with what appeared to be a library of knitting patterns. I explained that Leslie had been, as it were, accidentally gored by my horns. As usual, Mother looked upon the gloomy side and was convinced that I had secreted a bull in my room, which had disemboweled Leslie. Her relief at finding him sitting on the floor but apparently intact was considerable but tinged with annoyance.

"Leslie, dear, what have you been doing?" she asked.

Leslie gazed up at her, his face slowly taking on the color of a sun-mellowed plum. He had some difficulty in finding his voice.

"That bloody boy," he said at last, in a sort of muted roar. "He tried to brain me . . . hit me with a pair of sodding great deer horns!"

"Language, dear," said Mother automatically. "I'm sure he didn't mean to."

I said no, I had intended no harm, but in the interests of accuracy I would point out that they were not deer horns, which were a different shape, but the horns of a species of bull which I had not as yet identified.

"I don't care what bleeding species it is," snarled Leslie. "I don't care whether it's a bloody bastard brontosaurus horn!"

"Leslie, *dear*," said Mother, "it's quite unnecessary to swear so much."

"It *is* necessary," shouted Les, "and if you'd been hit on the head by something like a whale's rib cage, you'd swear too."

I started to explain that a whale's rib cage did not, in

fact, resemble my horns in the least, but I was quelled by a terrible look from Leslie, and my anatomical lecture dried in my throat.

"Well, dear, you can't keep them over the door," said Mother, "it's a most dangerous place. You might have hit Larry."

My blood ran cold at the thought of Larry felled by the horns of my bull.

"You'll have to hang them somewhere else," Mother continued.

"No," said Leslie. "If he must keep the bloody things, he's not to hang them up. Put them in a cupboard or somewhere."

Reluctantly I accepted this stricture, and so my horns reposed onto the windowsill, doing no further damage than to fall regularly onto our maid Lugaretzia's foot every evening when she closed the shutters, but as she was a professional hypochondriac of no mean abilities, she enjoyed the bruises she sustained. But this incident put a blight on my relationship with Leslie for some time, which was the direct cause of my unwittingly arousing Larry's ire.

Early in the spring I had heard echoing and booming from the reed beds around the salt pans the strange roaring of a bittern. I was wildly excited about this for I had never seen one of these birds and I was hopeful that they would nest, but pinpointing the exact area in which the birds were operating was difficult for the reed beds were extensive. However, by spending some considerable time perched in the higher branches of an olive tree on a hill commanding the reeds, I succeeded in narrowing down the field of search considerably. Soon the bitterns stopped calling, and I felt sure they were nesting. Having narrowed the area down to an acre or two, I set off early one morning, leaving the dogs behind. I soon reached the fields and plunged into the reed beds, moving to and fro like a questing hound, refusing to be tempted away from my objective by the sudden ripple of a water snake, the clop of a jumping frog or the tantalizing dance of a newly hatched butterfly. Soon I was in the heart of the cool, rustling reeds, and I then found, to my consternation, that the area was so extensive and the reeds were so high that I was completely lost. On every side I was surrounded by a fence of reeds, and their leaves made a flickering green canopy above me through which I could see the vivid blue sky. Being lost did not worry me, for I knew, if I walked long enough in any direction, I would hit the sea or the road; but what did worry me was that I could not be sure if I were searching the right area. I found some almonds in my pocket and sat down to eat them while I considered the problem.

I had just eaten the last one and decided that my best course was to go back to the olive trees and reestablish my bearings when I discovered that I had been sitting within eight feet of a bittern for the last five minutes without knowing it. He was standing there, stiff as a guardsman, his neck stretched up straight, his long, greenish-brown beak pointing skywards, while from each side of his narrow skull his dark, protuberant eyes gazed at me with a fierce watchfulness. His body, pale fawn mottled with dark brown, merged into the shimmering shadow-flecked reeds perfectly, and to add to the illusion that he was part of the moving background, the bird swayed from side to side. I was enchanted and sat watching him, hardly daring to breathe. Then there was a sudden commotion among the reeds, and the bittern abruptly stopped looking like a reed and launched himself heavily into the air as Roger, with lolling tongue and eyes beaming with bonhomie, came crashing into view.

I was torn between remonstrating with Roger for having frightened the bittern and praising him for his undoubted feat of having tracked me down by scent over a difficult route of about a mile and a half. However, Roger was obviously so delighted with his own achievement that I had not the heart to scold him; I found two almonds I had overlooked in my pocket and gave them to him as a reward. Then we set to work to search for the bitterns' nest. We soon found it, a neat pad of reeds with the first greenish egg lying in the cup. I was delighted and determined to keep a close watch on the nest to note the progress of the young; then, carefully bending the reeds to mark the trail, I followed Roger's stumpy tail. He obviously had a much better sense of direction than I had, for within a hundred yards we had reached the road and he was shaking the water off his woolly coat and rolling in the fine, dry, white dust.

As we left the road and made our way up the hillside through the olive groves sparkling with light and shade, colored with a hundred wild flowers, I stopped to pick some anemones for Mother, and while I gathered the wine-colored flowers I brooded on the problem of the bitterns. When the hen bird had reared her brood to the stage where they were fully feathered, I would dearly have liked to kidnap two and add them to my not inconsiderable menagerie of pets. The trouble was, the fish bill for my present creatures – a blackbacked gull, twenty-four terrapins, and eight water snakes – was considerable, and I felt that Mother would view the addition of two hungry young bitterns with mixed feelings, to say the least. Pondering this problem, it was some time before I became aware that someone was piping an urgent, beckoning call on a flute.

I glanced down at the road below and there was the rose-beetle man. He was a strange, itinerant peddler I frequently met on my expeditions through the olive groves. Slender, foxy-faced and dumb, he wore the most astonishing garb – a huge, floppy hat to which were pinned many strings tied to glittering goldy-green rose beetles, clothes mended with so many multicolored bits of cloth that it looked almost as though he were wearing a patchwork quilt, and a great, bright blue cravat to complete his ensemble. On his back were bags and boxes, cages full of pigeons, and from his pockets he could produce anything from wooden flutes, carved animals and combs to bits of the sacred robe of St. Spiridion. One of his chief charms as far as I was concerned was that, being dumb, he had to rely on his remarkable ability for mimicry, and he used his flute as his tongue. When he saw that he had caught my attention, he took the flute from his mouth and beckoned. I hurried down the hill for I knew that the rose-beetle man sometimes had things of great interest. It was he, for example, who had got me the biggest clam shell in my collection, and, moreover, with the two tiny parasitic pea-crabs still inside.

I stopped by him and said good morning. He smiled, showing discolored teeth, and doffed his floppy hat with an exaggerated bow, setting all the beetles that were tied to it buzzing sleepily on the ends of their strings like a flock of captive emeralds. Presently, after inquiring after my health by leaning forward and peering questioningly and anxiously, wide-eyed, into my face, he told me that he was well by playing a rapid, gay, rippling tune on his flute and then drawing in great lungfuls of warm spring air and exhaling them, his eyes closed in ecstasy. Thus having disposed of the courtesies, we got down to business.

What, I inquired, did he want of me? He raised his flute to his lips and gave a plaintive, quavering hoot, prolonged and mournful, and then, taking the flute from his lips, opened his eyes wide and hissed, swaying from side to side and occasionally snapping his teeth together. As an imitation of an angry owl it was so perfect that I almost expected the rose-beetle man to fly away. My heart beat with excitement, for I had long wanted a mate for my scops owl, Ulysses, who spent his days sitting like a carved totem of olive wood above my bedroom window and his nights decimating the mouse population around the villa. But the rose-beetle man laughed to scorn my idea of anything so common as a scops owl when I asked him. He removed a large cloth bag from the many bundles with which he was encumbered, opened it, and carefully tipped the contents out at my feet.

To say that I was struck speechless would be putting it mildly, for there in the white dust tumbled three huge owlets, hissing and swaying and beak-cracking in what seemed to be a parody of the rose-beetle man, their tangerine-golden eyes enormous with a mixture of rage and fear. They were baby eagle-owls, so rare as to be a prize almost beyond the dreams of avarice, and I knew that I must have them. The fact that the acquisition of the three fat and voracious owls would send the meat bill up in the same way that the addition of bitterns to my collection would affect the fish bill was neither here nor there. Bitterns were things of the future, which might or might not materialize, but the owls, like large, greyish-white snowballs, beak-clicking and rumbaing in the dust, were solid fact. I squatted down by the owlets, and as I stroked them into a state of semisomnolence, I bargained with the rose-beetle man. He was a good bargainer, which made the whole thing much more interesting, but it was also very peaceful bargaining with him for it was done in complete silence. We sat opposite one another like two great art connoisseurs at Christie's, say, bargaining over a trio of Rembrandts. The lift of a chin, the minutest inclination or half-shake of the head was sufficient, and there were long pauses during which the rose-beetle man tried to undermine my determination with the aid of music and some indigestible nougat which he had in his pocket. But it was a buyer's market and he knew it; who else in the length and breadth of the island would be mad enough to buy, not one, but three baby eagle-owls? Eventually the bargain was struck.

As I was temporarily embarrassed financially, I explained to the rose-beetle man that he would have to wait for payment until the beginning of the next month, when my pocket money was due. The rose-beetle man had frequently been in this predicament himself, so he understood the situation perfectly. I would, I explained, leave the money with our mutual friend Yani at the cafe on the crossroads, where the rosebeetle man could pick it up during one of his peregrinations across the countryside. Thus having dealt with the sordid, commercial side of the transaction, we shared a stone bottle of ginger beer from the rose-beetle man's capacious pack and then I placed my precious owls carefully in their bag and continued on my way home, leaving the rose-beetle man lying in the ditch among his wares and the spring flowers, playing on his flute.

It was the lusty cries that the owlets gave on the way back to the villa that suddenly brought home to me the culinary implications of my new acquisitions. It was obvious that the rose-beetle man had not fed his charges, and I did not know how long he had had them, but judging from the noise they were making, they were extremely hungry. It was a pity, I reflected, that my relations with Leslie were still slightly strained, for otherwise I could have persuaded him to shoot some sparrows or perhaps a rat or two for my new babies. As it was, I could see I would have to rely on my mother's unfailing kindness of heart.

I found her ensconced in the kitchen, stirring frantically at a huge, aromatical, bubbling cauldron, frowning at a cookbook in one hand, her spectacles misty, her lips moving silently as she read. I produced my owls with the air of one who is conferring a gift of inestimable value. My mother straightened her spectacles and glanced at the three hissing, swaying balls of down.

"Very nice, dear," she said in an absentminded tone of voice, "very nice. Put them somewhere safe, won't you?"

I said they would be incarcerated in my room and that nobody would know that I had got them.

"That's right," said Mother, glancing nervously at the owls. "You know how Larry feels about more pets."

I did indeed, and I intended to keep their arrival a secret from him at all costs. There was just one minor problem, I explained, and that was that the owls were hungry – were, in fact, starving to death.

"Poor little things," said Mother, her sympathies immediately aroused. "Give them some bread and milk."

I explained that owls ate meat and that I had used up the last of my meat supply. Had Mother perhaps a fragment of meat she could lend me so that the owls would not die?

"Well, I'm a bit short of meat," said Mother. "We've

having chops for lunch. Go and see what's in the icebox."

I went to the massive icebox in the larder that contained our perishable foodstuffs and peered into its icy, misty interior, but all I could unearth were the ten chops for lunch, and even these were hardly meal enough for three voracious eagle-owls. I went back with this news to the kitchen.

"Oh dear," said Mother. "Are you sure they won't eat bread and milk?"

I was adamant. Owls would only eat meat.

At that moment one of the babies swayed so violently he fell over, and I was quick to point this out to Mother as an example of how weak they were getting.

"Well, I suppose you'd better take the chops then," said Mother, harassed. "We'll just have to have vegetable curry for lunch."

Triumphantly I carried the owls and the chops to my bedroom and stuffed the hungry babies full of meat.

As a direct consequence of the owls' arrival, we sat down to lunch rather late.

"I'm sorry lunch is so late," said Mother, uncovering a tureen and letting loose a cloud of aromatic, curry-scented steam, "but the potatoes simply wouldn't cook for some reason."

"I thought we were going to have chops," said Larry aggrievedly. "I spent all morning getting my taste buds on tip-toe with the thought of chops. What happened to them?"

"I'm afraid it's the owls, dear," said Mother apologetically. "They have such huge appetites."

Larry paused, a spoonful of curry halfway to his mouth.

"Owls?" he said, staring at Mother. "Owls? What do you mean, owls? What owls?"

"Oh," said Mother, flustered, having realized that

she had made a tactical error, "just owls . . . birds, you know . . . nothing to worry about."

"Are we having a plague of owls?" Larry inquired. "Are they attacking the larder and zooming out with bunches of chops in their talons?"

"No, no, dear, they're only babies. They wouldn't do that. They have the most beautiful eyes," Mother said, "and they were simply starving, poor little things."

"Bet they're something new of Gerry's," said Leslie sourly. "I heard him crooning to something before lunch."

"Then he's got to release them," barked Larry.

I said I could not do this as they were babies.

"Only babies, dear," said Mother placatingly. "They can't help it."

"What do you mean, can't help it?" said Larry. "The damned things, stuffed to the gills with my chops – "

"*Our* chops," Margo interrupted. "I don't know why you have to be so selfish."

"It's got to stop," Larry went on, ignoring Margo. "You indulge the boy too much."

"They're just as much our chops as yours," said Margo.

"Nonsense, dear," said Mother to Larry, "you do exaggerate. After all, it's only some baby owls."

"Only!" said Larry bitingly. "He's already got one owl, and we know *that* to our cost."

"Ulysses is a very sweet bird and no trouble," said Mother defensively.

"Well, he might be sweet to you," said Larry, "but he hasn't come and vomited up all the bits of food he has no further use for all over *your* bed."

"That was a long time ago, dear," said Mother soothingly, "and he hasn't done it again."

"And what's it got to do with *our* chops, anyway?" asked Margo.

"It's not only owls," said Larry, "though, God knows, if this goes on, we'll start to look like Athene. You don't seem to have any control over him. Look at that business with the turtle last week."

"That was a mistake, dear," said Mother. "He didn't mean any harm."

"A mistake!" said Larry witheringly. "He disemboweled the bloody thing all over the verandah. My room smelt like the interior of Captain Ahab's boot. It's taken me a week and the expenditure of about five hundred gallons of eau de cologne to freshen it up to the extent where I can enter it without fainting."

"We smelt it just as much as you did," said Margo indignantly. "Anyone would think that you were the only one to smell it."

"Yes," said Leslie, "it smelt worse in my room. I had to sleep out on the back verandah. I don't know why you think you're the only one who ever suffers."

"I don't," said Larry. "I'm just not interested in the suffering of lesser beings."

"The trouble with you is you're selfish," said Margo, clinging to her original diagnosis.

"All right," snapped Larry, "don't listen to me. You'll all complain soon enough when your beds are waist-deep in owl vomit. I shall go and stay in a hotel."

"I think we've talked quite enough about the owls," said Mother firmly. "Who's going to be in for tea?"

It transpired that we were all going to be in for tea.

"I'm making some scones," said Mother, and sighs of satisfaction ran round the table, for Mother's scones, wearing cloaks of homemade strawberry jam, butter and cream, were a delicacy all of us adored. "Mrs. Vadrudakis is coming to tea, so I want you to behave," Mother went on.

Larry groaned.

"Who the hell is Mrs. Vadrudakis?" he asked. "Some old bore, I suppose?"

"Now, don't start," said Mother severely. "She sounds a very nice woman. She wrote me such a nice letter; she wants my advice."

"What on?" asked Larry.

"Well," said Mother, "she's very distressed by the way the peasants keep their animals. You know how thin the dogs and cats are, and those poor donkeys with sores that we see. Well, she wants to start a society for the elimination of cruelty to animals here in Corfu, rather like the RSPCA, you know. And she wants us to help her."

"She doesn't get my help," said Larry firmly. "I'm not helping any society to prevent cruelty to animals. I'd help them to promote cruelty."

"Now, Larry, don't say things like that," said Mother severely. "You know you don't mean it."

"Of course I do," said Larry, "and if this Vadrudakis woman spent a week in this house, she'd feel the same. She'd go round strangling owls with her bare hands, if only to survive."

"Well, I want you all to be polite," said Mother firmly, adding, "and you're not to mention owls, Larry. She might think we're peculiar."

"We are," said Larry with feeling.

After lunch I discovered that Larry, as he so often did by criticism, had alienated the two people who might have been his allies in his anti-owl campaign, Margo and Leslie. Margo, on seeing the owlets, went into raptures over them. She had just acquired the art of knitting, and with lavish generosity she offered to knit anything I wanted for the owls. I toyed with the idea of having them all dressed up in identical, striped pullovers but discarded this as impractical and reluctantly refused Margo's kind offer.

Leslie's offer of help was more concrete. He said he would shoot a supply of sparrows for me. I asked whether he could do this every day.

"Well, not every day," said Leslie. "I might not be here; I might be in town or something. But I will when I can."

I suggested that he might do some bulk shooting for me, procuring enough sparrows to last me a week, perhaps.

"That's a good idea," said Leslie, struck by this. "You work out how many you need for the week, and I'll get 'em for you."

Laboriously, for mathematics was not my strong point, I worked out how many sparrows (supplemented with meat) I would need a week and took the result to Leslie in his room, where he was cleaning the latest addition to his collection, a magnificent old Turkish muzzle-loader.

"Yes . . . OK," he said, looking at my additions, "I'll get 'em for you. I'd better use the air rifle; if I use the shotgun, we'll have bloody Larry complaining about the noise."

So, armed with the air rifle and a large paper bag, we went round the back of the villa. Leslie loaded the gun, leaned back against the trunk of an olive tree, and started shooting. It was really as simple as target shooting, for that year we had a plague of sparrows and the roof of the villa was thick with them. As they were picked off by Leslie's excellent marksmanship, they would roll down the roof and fall to the ground, where I would collect them and put them in my paper bag.

After the first few shots, the sparrows grew a little uneasy and retreated higher up until they were sitting on the apex of the roof. Here Leslie could still shoot them, but they were precipitated over the edge of the roof and rolled down to fall on the verandah on the other side of the house.

"Wait until I shoot a few more before you collect 'em," said Leslie, so I dutifully waited.

He continued shooting for some time, rarely missing, and the faint thunk of the rifle coincided with the collapse and disappearance of a sparrow from the rooftop.

"Damn," he said suddenly. "I've lost count. How many's that?"

I said that I hadn't been counting.

"Well, go and pick up the ones on the verandah and wait there. I'll pick off another six. That should do you."

Clasping my paper bag, I went around to the front of the house and saw, to my consternation, that Mrs. Vadrudakis, whom we had forgotten, had arrived for tea. She and Mother were sitting somewhat stiffly on the verandah clasping cups of tea, surrounded by the bloodstained corpses of numerous sparrows.

"Yes," Mother was saying, obviously hoping that Mrs. Vadrudakis had not noticed the rain of dead birds, "yes, we're all great animal lovers."

"I hear this," said Mrs. Vadrudakis, smiling benevolently. "I hear you lof the animals like me."

"Oh, yes," said Mother. "We keep so many pets. Animals are a sort of passion with us, you know."

She smiled nervously at Mrs. Vadrudakis, and at that moment a dead sparrow fell into the strawberry jam.

It was impossible to cover it up and equally impossible to pretend it was not there. Mother stared at it as though hypnotized; at last, she moistened her lips and smiled at Mrs. Vadrudakis, who was sitting with her cup poised, a look of horror on her face.

"A sparrow," Mother pointed out weakly. "They . . . er . . . seem to be dying a lot this year."

At that moment Leslie, carrying the air rifle, strode out of the house.

"Have I killed enough?" he inquired.

The next ten minutes were fraught with emotion. Mrs. Vadrudakis said she had never been so insulted in her life and that we were all fiends in human shape. Mother kept saying that she was sure Leslie had not meant it as an insult and that, anyway, she was sure the sparrows had not suffered. Leslie, loudly and belligerently, kept saying it was a lot of bloody fuss about nothing, and anyway, owls ate sparrows and did Mrs. Vadrudakis want the owls to starve, eh? But Mrs. Vadrudakis refused to be comforted. She wrapped herself, a tragic and insulted figure, in her cloak, shudderingly picked her way through the sparrows' corpses, got into her cab and was driven away through the olive groves at a brisk trot.

"I do wish you children wouldn't do things like that," said Mother, shakily pouring herself a cup of tea while I picked up the sparrows. "It really was most . . . careless of you, Leslie."

"Well, how was I to know the old fool was out here?" said Leslie indignantly. "I can't be expected to see through the house, can I?"

"You should be more careful, dear, said Mother. "Heaven knows what she must think of us."

"She thinks we're savages," said Leslie, chuckling. "She said so. She's no loss, silly old fool."

"Well, the whole thing's given me a headache," said Mother. "Go and ask Lugaretzia to make some more tea, Gerry, will you?"

Two pots of tea and several aspirins later, Mother was beginning to feel better. I was sitting on the verandah giving her a lecture on owls, to which she was only half-listening, saying, "Yes, dear, how interesting," at intervals, when she was suddenly galvanized by a roar of rage from inside the villa.

"Oh dear, I can't stand it," she moaned. "Now what's the matter?"

Larry strode out onto the verandah.

"Mother!" he shouted. "This has got to stop. I won't put up with it."

"Now, now, dear, don't shout. What's the matter?" Mother inquired.

"It's like living in a bloody natural history museum!"

"What is, dear?"

"This is! Life here. It's intolerable. I won't put up with it!" shouted Larry.

"But what's the matter, dear?" Mother asked, bewildered.

"I go to get myself a drink from the icebox and what do I find?"

"What do you find, dear?" asked Mother with interest.

"Sparrows!" bellowed Larry. "Bloody great suppurating unhygienic bags of sparrows!"

It was not my day.

Fakirs and Fiestas

The prince of darkness is a gentleman.
SHAKESPEARE, *King Lear*

IT WAS ALWAYS during the late spring that my collection of animals swelled to a point where even Mother occasionally grew alarmed, for it was then that everything was arriving and hatching, and baby animals are, after all, easier to acquire than adults. It was also the time when the birds, newly arrived to nest and rear their young, were harried by the local gentry with guns in spite of the fact that it was out of season. Everything was grist to their mill, these townee sportsmen, for whereas the peasants would stick to the so-called game birds – thrushes, blackbirds and the like – the hunters from the town would blast everything that flew, and you would see them returning triumphantly weighed down with guns and bandoliers of cartridges, their game bags full of a sticky, bloody, feathery conglomeration of anything from robins to redstarts, from nuthatches to nightingales. So in the spring my room and that portion of the verandah set aside for the purpose always had at least half a dozen cages and boxes containing gape-mouthed baby birds or birds that I had managed to rescue from the sportsmen and that were recuperating with makeshift splints on wings or legs.

The only good thing about this spring slaughter was

that it gave me a pretty good idea of what birds were found in the island, for finding I could not stop the killing, I could at least turn it to good account, and I would track down the brave and noble Nimrods and ask to see the contents of their game bags. I would make a list of all the dead birds and, by pleading, save the lives of those that had only been wounded. It was by this means that Hiawatha came into my possession.

I had spent an interesting and energetic morning with the dogs. We had been up early and out in the olive groves while everything was still dawn-chilly and misted with dew; I had found this an excellent time for collecting insects, for the coldness made them lethargic and unwilling to fly. I had obtained two butterflies and a moth new to my collection, two unknown beetles, and seventeen locusts, which I collected to feed my baby birds with. By the time the sun was well up in the sky and had gathered some heat, we had unsuccessfully chased a snake and a green lizard, milked Agathi's goat (unbeknownst to her) into a collecting jar as we were all thirsty, and dropped in on my old shepherd friend Yani, who provided us with some bread and fig cake and a straw hat full of wild strawberries to sustain us.

We made our way down to a small bay, where the dogs lay panting or crab-hunted in the shallows while I, spread-eagled like a bird in the warm, transparent water, lay face downwards holding my breath and drifting over the landscape of the sea. When it grew close to midday and my stomach told me lunch would be ready, I dried off in the sun, the salt forming in patches on my skin like a silky pattern of delicate lace, and started off home. As we meandered through the olive groves, shady and cool as a well between the great trunks, I heard a series of explosions away to the right in the myrtle groves. I moved over to investigate, keeping the dogs close to me, for Greek hunters were jumpy and would in most cases shoot before stopping to identify what they were shooting at. The danger applied to me too, so I talked loudly to the dogs as a precaution. "Here, Roger, heel! Good boy. Puke! Widdle! Widdle, come here! Heel, that's a good boy. Puke, come back." I spotted the hunter sitting on a giant olive root mopping his brow, and as soon as I knew he had seen us, I approached him.

He was a plump, white little man, with a moustache like an elongated black toothbrush over his prim little mouth, and dark glasses covered eyes as round and as liquid as a bird's. He was dressed in the height of fashion for hunting – polished riding boots, new breeches in white cord, an atrociously cut hacking jacket in mustard and green tweed, beset with so many pockets it looked like the eaves of a house hung with swallows' nests. His green Tyrolean hat, with its bunch of scarlet and orange feathers, was tilted to the back of his curly head, and he was mopping his ivory brow with a large handkerchief that smelled strongly of cheap cologne.

"*Kalimera, kalimera,*" he greeted me, beaming and puffing. "Welcome. Houf! it's a hot day, isn't it?"

I agreed and offered him some of the strawberries that remained in my hat. He looked at them rather apprehensively, as if fearing they were poisoned, took one delicately in his plump fingers and smiled his thanks nervously as he popped it into his mouth. I got the strong impression that he had never eaten strawberries out of a hat with his fingers and was not quite sure about the rules.

"I've had a good morning's hunt," he said proudly, pointing to where his game bag lay, bulging ominously, blood-bespattered and feathery. From the mouth of it protruded the wing and head of a lark, so blasted and mangled it was difficult to identify.

Would he, I inquired, mind if I examined the contents of his bag.

"No, no, of course not," he said. "You will see I'm quite a marksman."

I did see. His bag consisted of four blackbirds, a golden oriole, two thrushes, eight larks, fourteen sparrows, two robins, a stonechat, and a wren. The last, he admitted, was a bit small but very sweet to eat if cooked with paprika and garlic.

"But this," he said proudly, "is the best. Be careful, because it's not quite dead."

He handed me a bloodstained handkerchief, and I unwrapped it carefully. Inside, gasping and exhausted, a great hard seal of blood on its wing, was a hoopoe.

"That is not, of course, good to eat," he explained to me, "but the feathers will look good in my hat."

I had long wanted to possess one of these splendid, heraldic-looking birds, with their fine crests and their salmon-pink and black bodies, and I had searched everywhere for their nests so that I could hand-rear some young ones. Now here was a live hoopoe in my hands, or, to be more exact, a half-dead one. I examined it carefully and found that it in fact looked worse than it was, for all it had was a broken wing, and this was a clean break as far as I could judge. The problem was how to get my proud, fat hunter to part with it.

Suddenly I had an inspiration. I started by saying that it made me feel bitter and annoyed that my mother was not there at that very moment, for she was, I explained, a world-famous authority on birds. (Mother could, with difficulty, distinguish between a sparrow and an ostrich.) She had, in fact, I went on, written the definitive work on birds for the hunters of England. To prove it, I produced from my collecting bag a much-battered and much-consulted copy of *A Bird Book for the Pocket* by Edmund Sanders, a book I was never without. My fat friend was most impressed; he turned over the pages muttering appreciative *po po po pos* to himself. My mother, he said, must be a remarkable woman to have written such a book. The reason I wished she was there at this moment, I went on, was because she had never seen a hoopoe. She had seen every other bird on the island, including the rare kingfisher; to prove it I took the scalp of a dead kingfisher I had found and used as a talisman from my collecting bag and laid it in front of him. He was much struck with this little skullcap of bright blue feathers. They were, I said, much prettier than hoopoe feathers when one considered it. It took a little time for the thought to penetrate, but I soon had him begging that I would take the hoopoe to my mother in exchange for the scrap of velvety blue feathers. I put on a nice display of astonished reluctance fading into groveling gratitude, put the wounded hoopoe inside my shirt and hurried home with it, leaving my hunter friend sitting on his olive root looking like Tweedledum, trying happily to fix the kingfisher scalp to his hat with a pin.

When I got home, I took my new acquisition to my room and examined it carefully. To my relief, its long, curved, rubbery beak, like a slender scimitar, was intact, for without the use of this delicate organ, I knew that the bird could not survive. Apart from exhaustion and fright, the only thing wrong with it appeared to be a broken wing. The break was high in the upper wing, and on investigating it gently, I found that it was a clean break, the bone having been snapped like a dry twig, not smashed and splintered like a green one. I carefully cut away the feathers with my dissecting scissors, washed the scab of blood and feathers away with warm water and disinfectant, splinted the bone with two curved slivers of bamboo, and bound the whole thing up tight. It was quite a professional job, and I was proud of it. The only trouble was that it was too heavy. When I released the bird, it fell over on its side, dragged down by the weight of the splint. After some experiment, I managed to make a much lighter splint out of bamboo and sticking plaster, and with a thin strip of bandage bound the whole thing firmly to the bird's side. Then, with a pipette, I gave it a drink of water and placed it in a cardboard box covered with a cloth to recover.

I called the hoopoe Hiawatha, and the family greeted its arrival in our midst with unqualified approval, for they all liked hoopoes and, moreover, it was the only exotic bird species they could all recognize at twenty paces. Finding food for Hiawatha kept me very busy during the first few days of her convalescence, for she was a finicky patient, would eat only live food and was very choosy about that. I had to release her on the floor of my room and throw the tidbits at her – the succulent grasshoppers, as green as jade, locusts with plump thighs, their wings as crisp as biscuits, small lizards and tiny frogs. These she would grab and bang vigorously on any suitable hard surface – a chair or bed leg, the edge of the door or table – until she was sure that they were dead. Then, a couple of quick gulps and she would be ready for the next course. One day, when the family had all assembled in my room to watch Hiawatha feed, I gave her an eight-inch slowworm. With her delicate beak, her finely banded crest, and her beautiful pink and black color scheme, she looked a very demure bird, even more so because she generally kept her crest folded back against her skull. But now she took one look at the slowworrn and changed into a predatory monster. Her crest rose and spread itself, quivering like a peacock's tail, her throat puffed out, and she uttered a strange little purring grunt deep in her throat and hopped rapidly and purposefully towards where the slowworm was dragging its burnished-copper body along, oblivious of its fate. Hia-watha paused, and with her splinted and her good wing spread out, she leaned forward and pecked at the slowworm – a rapid rapier thrust of her beak, so quick it was difficult to see. The slowworm, at the blow, writhed into a lashing figure of eight, and I saw to my amazement that Hiawatha's first blow had completely crushed the reptile's eggshell-fragile skull.

"Good Lord!" said Larry, equally amazed. "Now, that's what I call a *useful* bird to have around the house. A few dozen of those around and we wouldn't have to worry about snakes."

"I don't think they could tackle a big one," said Leslie judiciously.

"Well, I wouldn't mind if they just cleaned up the small ones," said Larry. "That'd be a start."

"You talk as if the house were full of snakes, dear," said Mother.

"It is," said Larry austerely. "What about the Medusa wig of snakes Leslie found in the bath?"

"They were only water snakes," said Mother.

"I don't care what they were," said Larry. "If Gerry's going to be allowed to fill the bath with snakes, then I shall carry a brace of hoopoes around with me."

"Ooh, look at it now!" squeaked Margo.

Hiawatha had delivered a number of rapid blows down the length of the slowworm's body, and she was now picking up the still writhing length and dashing it onto the floor rhythmically, as the fishermen would beat an octopus against the rocks to make it tender. After a time there was no discernible life left in the body; Hiawatha stared down at it, crest up, head on one side. Satisfied, she seized the head in her beak and slowly, gulping and throwing her head back, she swallowed it inch by inch. In a couple of minutes there was only half an inch of tail protruding from the corner of her beak.

Hiawatha never grew really tame and she was always nervous, but she learned to tolerate human beings in fairly close proximity to her. When she had settled down, I used to take her out onto the verandah where I kept my various other birds and let her walk about in the shade of the grapevine. It was not unlike a hospital ward, for at that time I had six sparrows recovering from concussion brought about by being caught in break-back mousetraps set by peasant boys, four blackbirds and a thrush who had been caught by baited fishhooks set in the olive groves by peasant boys, and half a dozen assorted birds ranging from a tern to a magpie recovering from the effects of gunshot wounds caused by the parents of the peasant boys. In addition there were a nest of young goldfinches and an almost fledged green finch which I was hand-rearing. Hiawatha did not seem to mind the proximity of these other birds but she kept to herself, pacing slowly up and down the flagstones, brooding, with half-closed eyes, aloofly aristocratic like a beautiful queen imprisoned in some castle. At the sight of a worm, frog or grasshopper, of course, her behavior would become anything but queenly.

About a week after Hiawatha had entered my avian clinic, I set off one morning to meet Spiro. This was a sort of daily ritual; he would blow loud blasts on his horn when he reached the edge of the property, which was some fifty acres in extent, and the dogs and I would tear through the olive groves to intercept him somewhere along the drive. Panting for breath, I would burst out of the olive groves, the dogs barking hysterically in front of me, and we would hold up the great, gleaming Dodge, its hood back, Spiro in his peaked cap crouching, massive, brown and scowling behind the wheel. I would take my place on the running board, holding tight to the windshield, and Spiro would drive on, the dogs in an ecstasy of mock fierceness trying to bite the front tires. The conversation every morning was also a ritual that never varied.

"Goods mornings, Master Gerrys," Spiro would say. "Hows are yous?"

Having ascertained that I had not developed any dangerous disease during the night, he would inquire after the rest of us.

"And hows the familys?" he would ask. "Hows your mothers? And Master Larrys? And Master Leslies? And Missy Margos?"

By the time I had reassured him as to their health, we would have reached the villa, where he would lumber from one member of the family to the other, checking as to whether my information was correct. I was rather bored by the daily, almost journalistic interest Spiro took in the family's health, as if they were royalty, but he persisted in it as if some awful fate might have overtaken them during the night. One day, in a fit of devilry, I had told him, in response to his earnest inquiry, that they were all dead; the car swerved off the drive and crashed straight into a large oleander bush, showering Spiro and me with pink blossoms and nearly knocking me off the running board.

"Gollys, Master Gerrys! You mustn't says things like thats!" he roared, pounding the wheel with his fist.

"You makes me scarce when you say things likes that. You makes me sweats! Don't you ever say that agains."

This particular morning, having reassured himself as to the health of each member of the family, he lifted a small strawberry basket covered with a fig leaf from the seat by his side.

"Here," he said, scowling at me. "I gots a presents for you."

I took the leaf off the basket, inside which crouched two naked and repulsive-looking birds. I was enchanted and thanked Spiro profusely, for they were baby jays, as I could see by their sprouting wing feathers, and I had never had jays before. I was so pleased with them that I took them with me when I went to my studies with Mr. Kralefsky. This was the advantage of having a tutor who was as mad about birds as I was. Together we spent an exciting and interesting morning trying to teach them to open their mouths and feed, when we should have been committing the glittering pageantry of English history to memory. But the babies were singularly stupid and refused to accept either Kralefsky or me as a substitute mother. I took them back home at lunchtime and during the afternoon tried to get them to behave sensibly, but without success. They would only take food if I forced their beaks open and pushed it down their throats with my finger, a process that they strongly objected to, as well they might. Eventually, having shoved enough down them to more or less keep them alive, I left them in their strawberry basket on the verandah and went to fetch Hiawatha, who had shown a marked preference for having her food served on the verandah rather than in the privacy of my room. I placed her on the flagstones and started to throw her the grasshoppers I had caught for her. She hopped forward eagerly, snapped up the first, killed it, and swallowed it with almost indecent haste. As she sat there gulping, looking rather like an elderly, angular dowager duchess who had swallowed a sorbet the wrong way at a ball, the two baby jays, lolling their heads, bleary-eyed, over the edge of their basket caught sight of her. Immediately, they started to call wheezily, open-mouthed, their heads wobbling from side to side like two very old men looking over a fence. Hiawatha put up her crest and stared at them. I did not expect her to take very much notice, for she always ignored the other baby birds I had when they called out to be fed, but she hopped nearer the basket and surveyed the baby jays interestedly. I threw her a grasshopper and she grabbed it, killed it, and then, to my complete astonishment, hopped up to the basket and crammed the insect down the gaping maw of one of the jays. Both babies wheezed and screamed and flapped their wings in delight, and Hiawatha looked as startled at what she had done as I was. I threw her another grasshopper and she killed it and fed the other baby. After this, I would feed Hiawatha in my room and then bring her down onto the verandah periodically and she would act the part of mother to the baby jays.

She never showed any other maternal feelings for the babies; she would not, for example, seize the little encapsulated blobs of excreta from the babies' behinds when they cocked them over the edge of the nest. This task of cleaning up was left to me. Once she had fed the babies so that they stopped screaming, she lost all interest in them. I concluded it must be something in the timbre of their call that aroused her maternal instincts, for although I experimented with the other babies I possessed and they screamed their lungs out, she took no notice at all. Gradually the baby jays decided to let me feed them, and as soon as they stopped calling at her appearance, Hiawatha took no further notice of them. It was not simply that she ignored them; she seemed unaware of their existence.

When her wing had healed, I removed the splint and found that although the bone had set well, the wing muscles had become weak with lack of use, and Hiawatha tended to favor the wing, always walking rather than flying. To make her exercise it, I used to take her down into the olive groves and throw her up into the air so that she was forced to use her wings to make a safe landing. Gradually she started to take short flights herself as the wings strengthened, and I began to think that I would be able to release her, when she met her death.

I had taken her out onto the verandah one day, and while I was feeding my assortment of babies, Hiawatha flew or, rather, glided down to a nearby olive grove to practice her flying and make a light snack of some daddy longlegs that were just hatching. I was absorbed in feeding the babies, so I was not taking much notice, when suddenly I heard desperate, hoarse, despairing cries from Hiawatha. I vaulted over the verandah rail and raced through the trees, but I was too late. A large, mangy, battle-scarred feral cat was standing with the limp form of the hoopoe in his mouth, his great green eyes staring at me over her pink body. I gave a shout and ran forward; the cat turned with oillike fluidity and leapt into the myrtle bushes, carrying Hiawatha's body with him. I gave chase, but once the cat had reached the tangled sanctuary of the myrtles it was impossible to track him down. I returned, furious and upset, to the olive grove, where all that was left to remind me of Hiawatha were some pink feathers and a few drops of blood scattered like rubies on the grass. I swore that if I ever came across the cat again, I would kill it if I could. Apart from anything else, it represented a threat to the rest of my bird collection.

My mourning for Hiawatha was cut short by the arrival in our midst of something slightly more exotic than a hoopoe and much more trouble. Larry had suddenly announced that he was going to stay with some friends of his in Athens and do some research work. After the flurry of his departure, tranquility descended on the villa. Leslie spent most of his time pottering about with a gun, and Margo, who at that moment was not engaged in any hectic affair of the heart, had taken up soap sculpture. Ensconced in the attic, she was producing somewhat lopsided and slippery pieces of sculpture out of an acrid-smelling yellow soap and appearing in a flowered smock and an artistic trance at mealtimes. Mother, seizing on this unexpected period of calm, decided to do a job that had long wanted doing. The previous year had been an exceptionally good one for fruit, and Mother had spent hours preparing various jams and chutneys, some from her grandmother's recipes from India dating back to the early eighteen hundreds. Everything went fine, and the big cool larder was aglint with battalions of bottles. Unfortunately, during a particularly savage storm in the winter the larder roof had leaked, and in consequence Mother had come down one morning and found that all the labels had come off. She was now faced with several hundred jars, the contents of which were difficult to identify unless you opened them. Now, given a moment's respite by her family, she determined to do this necessary job. Since it involved tasting, I offered to help, and so between us we had got some hundred and fifty jars of preserves on the kitchen table, armed ourselves with spoons and new labels, and were just about to start on the mammoth tasting when Spiro arrived.

"Goods afternoons, Mrs. Durrells. Goods afternoons, Master Gerrys," he rumbled, lumbering into the kitchen like a chestnut-brown dinosaur. "I's gots a telegrams for you, Mrs. Durrells."

"A telegram, Spiro?" Mother quavered. "Who from, I wonder? I hope it's not bad news."

"No, don't you worrys, it's not bad news, Mrs. Durrells," he said, handing her the telegram. "I gots the man in the post office to reads it to me. It's from Master Larrys."

"Oh dear," said Mother with foreboding.

The telegram said simply, "Forgot to tell you Prince Jeejeebuoy arriving eleventh short stay. Athens wonderful. Love. Larry."

"Really, Larry is the most annoying creature!" Mother exclaimed angrily. "What does he go and invite a prince for? He knows we haven't got the right rooms for royalty. And he won't be here to entertain him. What am I supposed to do with a prince?"

She gazed at us in a distraught fashion, but neither Spiro nor I could give her any intelligent advice. We could not even telegraph Larry and demand his return, for, characteristically, he had gone off and omitted to give us his friends' address.

"The eleventh is tomorrow, isn't it? He'll be coming on the boat from Brindisi, I expect. Spiro, would you meet him and bring him out? And will you bring some lamb for lunch? Gerry, go and tell Margo to put some flowers in the spare room and to make sure the dogs haven't put any fleas in there, and tell Leslie he must go down to the village and tell Red Spiro we want some fish. Really, it's too bad of Larry. I shall give him a piece of my mind when he gets back. I can't be bothered with entertaining princes at my age," said Mother, bustling angrily and aimlessly around the kitchen, banging saucepans and frying pans about.

"I'll brings you some of my dahlias for the tables. Do you wants any champagne?" asked Spiro, who obviously felt that the prince should be treated properly.

"No. If he thinks I'm paying a pound a bottle for champagne, he's mistaken. He can just drink ouzo and wine like the rest of us, prince or no prince," said Mother firmly, and then added, "Well, I suppose you'd better bring a crate. We needn't give him any, and it always comes in useful."

"Don'ts you worrys, Mrs. Durrells," said Spiro comfortingly, "I fixes anythings you wants. You wants I gets the king's butler again?"

The king's butler, an ancient and aristocratic old boy, was dragged out of retirement by Spiro every time we had a big party.

"No, no, Spiro, we're not going to go to a lot of trouble. After all, he's coming unexpectedly so he'll just have to take us as he finds us. He'll just have to take potluck . . . and . . . and . . . muck in. And if he doesn't like it . . . well, it's just too bad," said Mother, shelling peas with trembling hands and dropping more on the floor than into the colander. "And, Gerry, go and ask Margo if she could run up those new curtains for the dining room. The material's in my bedroom. The old ones don't look the same since Les set fire to them."

So the villa was transformed into a hive of activity. The wooden floor of the spare room was scrubbed until it was a pale cream color, just in case the dogs had put any fleas in there; Margo ran up the new curtains in record time and did flower arrangements everywhere;

and Leslie cleaned his guns and boat in case the prince should want to go shooting or yachting. Mother, scarlet with heat, trotted frantically around the kitchen making scones, cakes, apple turnovers and brandy snaps, stews, pies, jellies and trifles. I was merely told to remove all my animals from the verandah and to keep them under control, to go and have my hair cut and to make sure I put on a clean shirt. So the following day, all dressed up by Mother's orders, we sat on the verandah and waited patiently for the prince to be brought out to us by Spiro.

"What's he a prince *of*?" asked Leslie.

"Well, I don't really know," said Mother. "One of those small states the maharajas have, I expect."

"It's a very odd name, Jeejeebuoy," said Margo. "Are you sure it's real?"

"Of course it's real, dear," said Mother. "There are lots of Jeejeebuoys in India. It's a very old family, like . . . um . . . like. . . "

"Smith?" suggested Leslie.

"No, no, not nearly as common as that. No, the Jeejeebuoys go right back in history," said Mother. "There must have been Jeejeebuoys long before my grandparents went to India."

"His ancestors probably organized the mutiny," said Leslie with relish. "Let's ask him if his grandfather invented the Black Hole of Calcutta."

"Oh, yes, let's," said Margo. "D'you think he did? What was it?"

"Leslie, dear, you shouldn't say things like that," said Mother. "After all, we must forgive and forget."

"Forgive and forget what?" asked Leslie, bewildered, not having followed Mother's train of thought.

"Everything," said Mother firmly, adding, rather obscurely, "I'm sure they *meant* well."

Before Leslie could investigate this further, the car roared up the drive and drew up below the verandah with an impressive squeal of brakes. Sitting in the back, dressed in black, with a beautifully arranged turban as white as a snowdrop bud, sat a slender, diminutive Indian with enormous, glittering, almond-shaped eyes that were like pools of liquid agate fringed with eyelashes as thick as a carpet. He opened the door deftly and leapt out of the car. His smile of welcome was like a lightning flash of white in his brown face.

"Vell, vell, here ve are at last," he cried excitedly, spreading his slender brown hands like butterfly wings and dancing up onto the verandah. "You must be Mrs. Durrell, of course. Such charm. And you are the hunter of the family . . . Leslie. And Margo, the beauty of the island, vithout doubt. And Gerry, the savant, the naturalist *par excellence*. I can't tell you how hot it makes me to meet you all."

"Oh . . . well . . . er . . . er . . . yes, we're delighted to meet you, Your Highness," Mother began.

Jeejeebuoy uttered a yelp and slapped his forehead.

"Desh and demnation!" he said. "My foolish name again! My dear Mrs. Durrell, how can I apologize? Prince is my Christian name. A vhim on my mother's part to make our humble family royal, you understand? A mother's love, hm? Dream son vill aspire to golden heights, huh? No, no, poor voman, ye must forgive her, uh? I am plain Prince Jeejeebuoy, at your service."

"Oh," said Mother, who, having geared herself to cope with royalty, felt somewhat let down. "Well, what do we call you?"

"My friends, of which I have an inordinate number," said the new arrival earnestly, "call me Jeejee. I do hope that you vill call me the same."

So Jeejee took up residence, and during the short time he was there he created more havoc and endeared himself more to us than any other guest we had had. With his pedantic English and his earnest, gentle air, he took such a deep and genuine interest in everything and everyone that he was irresistible. For Lugaretzia he had various pots of evil-smelling sticky substances with which to anoint her numerous imaginary aches and pains; with Leslie he would discuss in grave detail the state of hunting in the world and give graphic and probably untrue stories of tiger and wild boar hunts he had been on. For Margo he procured some lengths of cloth and made them into saris and taught her how to wear them; Spiro he would enthrall with tales of the riches and mysteriousness of the East, tales of bejeweled elephants wrestling with each other and maharajas worth their weight in precious stones. He was proficient with his pencil and, as well as taking a deep and genuine interest in all my pets, he completely won me over by doing delicate little sketches of them for me to stick in my natural history diary, a document which was, to my mind, considerably more important than a combination of Magna Carta, the Book of Kells and the Gutenberg Bible, and was treated as such by our discerning guest. But it was Mother that Jeejee really charmed into submission, for not only did he have endless mouth-watering recipes for her to write down and a fund of folklore and ghost stories, but his visit enabled Mother to talk endlessly about India, where she had been born and bred and which she considered her real home.

In the evening we would sit long over our meal at the big, creaking dining table, the clusters of oil lamps in the corners of the great room blooming in pools of primrose-yellow light, the drifts of small moths fluttering against them like snow. The dogs lying in the doorway – now that their numbers had risen to four, they were never allowed into the dining room – would yawn and sigh at our tardiness, but we would be oblivious to them. Outside, the ringing cries of the crickets and the crackle of tree frogs would make the velvety night alive. In the lamplight Jeejee's eyes would seem to grow bigger and blacker, like an owl's, with a strange, liquid fire in them.

"Of course, in your day, Mrs. Durrell, things vere different. You could not intermingle. No, no, strict segregation, vasn't it? But now things are better. First the maharajas got their toes in the doors, and nowadays even some of us humbler Indians are allowed to intermingle and thus accrue some of the advantages of civilization," said Jeejee one evening.

"In my day," said Mother, "it was the Eurasians that they felt most strongly about. We wouldn't be allowed to even play with them by my grandmother. Of course, we always did."

"Children are singularly insensitive to the correct civilized behavior," said Jeejee, smiling. "Still, there vere some difficulties at first, you know. Rome, however, vas not built in a day. Did you hear about the babu in my town who vas invited to the ball?"

"No," said Mother, "what happened?"

"Vell, he saw that after the gentlemen had finished dancing with the ladies, they escorted them back to their chairs and fanned them with the ladies' fan. So, having conducted a sprightly valtz with a European lady of some eminence, he conducted her safely back to her seat, took her fan, and said, 'Madam, may I make vind in your face?'"

"That sounds the sort of thing Spiro would say," said Leslie.

"I remember once," said Mother, throwing herself into reminiscence with pleasure, "it was when my husband was chief engineer in Rourki. We had the most terrible cyclone, Larry was only a baby. The house was a long, low one, and I remember we ran from room to room trying to hold the doors shut against the wind. As we ran from room to room, the house simply collapsed behind us. We eventually ended up in the butler's pantry. But when we had the house repaired, the babu contractor sent in a bill which was headed, 'For repairs to chief engineer's backside.'"

"India must have been fascinating then," said Jeejeebuoy, "because, unlike most Europeans, you vere part of the country."

"Oh, yes," said Mother, "even my grandmother was born there. When most people talked of home, they meant England; when we said home, we meant India."

"You must have traveled extensively," said Jeejee enviously. "I suppose you've seen more of my country than I have."

"Practically every nook and cranny," said Mother. "My husband being a civil engineer, of course, he had to travel. I always used to go with him. If he had to build a bridge or a railway right out in the jungle, I'd go with him and we'd camp."

"That must have been fun," said Leslie enthusiastically, "a primitive life under canvas."

"Oh, it was. I loved the simple life in camp," said Mother. "I remember the elephants used to go ahead with the marquees, the carpets and the furniture, and then the servants would go in the oxcarts with the linen and silver."

"You call that camping?" interrupted Leslie incredulously, "with marquees?"

"We only had three," said Mother defensively. "A bedroom, dining room and a drawing room. And they were built with fitted carpets anyway."

"Well, I don't call that camping," said Leslie.

"It was," said Mother. "It was right out in the jungle. We could hear tigers, and all the servants were terrified. Once they killed a cobra under the dining table."

"And that was before Gerry was born," said Margo.

"You should write your memoirs, Mrs. Durrell," said Jeejee gravely.

"Oh, no," laughed Mother, "I couldn't possibly write. Besides, what would I call it?"

"How about *It Took Fourteen Elephants*?" suggested Leslie.

"Or *Through the Forest on a Fitted Carpet*," suggested Jeejee.

"The trouble with you boys is you never take anything seriously," said Mother severely.

"Yes," said Margo, "I think it was jolly brave of Mother to camp with only three marquees and cobras and things."

"Camping!" snorted Leslie derisively.

"Well, it was camping, dear. I remember once one of the elephants went astray and we had no clean sheets for three days. Your father was most annoyed."

"I never knew anything as big as an elephant could go astray," said Jeejee, surprised.

"Oh, yes," said Leslie, "easily mislaid, elephants."

"Well, anyway, you wouldn't like it if you were without clean sheets," said Mother with dignity.

"Of course they wouldn't," said Margo defensively, "and I think it's fun hearing about ancient India, even if they don't."

"But I do find it most educational," Jeejee protested.

"You're always making fun of Mother," said Margo,

"and I don't see why you should be so superior just because your father invented the Black Hole or whatever it was."

It says much for Jeejee that he almost fell under the table laughing, and all the dogs started barking vociferously at his mirth.

But probably the most endearing thing about Jeejee was his intense enthusiasm for anything he happened to take up, even when it was demonstrated beyond a shadow of a doubt that he could not achieve success in that sphere of activity. When Larry had first met him, he had decided to be one of India's greatest poets, and with the aid of a compatriot who spoke little English ("He vas my compositor," Jeejee explained), he started a magazine called *Poetry for the People*, or *Potry for the Peeple* or *Potery for the Peopeople*, depending on whether Jeejee was supervising his compositor or not. This little magazine was published once a month, with contributions from everyone that Jeejee knew, and some of them made strange reading, as we discovered, for Jeejee's luggage was full of blurred copies of his magazine, which he would hand out in vast quantities to anyone who displayed interest. Perusing them, we discovered such interesting items as "The Potry of Stiffen Splendour – a creetical evaluation." Jeejee's compositor friend apparently believed in printing words as they sounded, or, rather, as they sounded to him at that moment. Thus there was a long and eulogistic article by Jeejee on "Tees Ellyot, Pot Supreme." The compositor's novel spelling, combined with the misprints naturally to be found in such a work, made reading it a pleasurable though puzzling occupation. "Wye Notte a Black Pot Lorat?" for example, posed an almost unanswerable question, written apparently in Chaucerian English; while the article entitled "Roy Cambill, Ball Fighter and Pot" made one wonder what poetry was coming to. However, Jeejee was undaunted by the difficulties, including the fact that his compositor never pronounced the letter *h* and so never used it. His latest enthusiasm was to start a second magazine (printed on the same hand press with the same carefree compositor), devoted to his newly evolved study of what he called "Fakyo," which was described in the first copy of *Fakyo for All* as "an amalgum of the misterious East, bringing together the best of Yoga and Fakirism, giving details and tiching people ow."

Mother was greatly intrigued by Fakyo, until Jeejee started to practice it. Clad in a loincloth and covered with ashes, he meditated for hours on the verandah or else walked in a well-simulated trance through the house, leaving a trail of ashes behind him. He fasted religiously for four days, and on the fifth day worried Mother to death by fainting and falling down the stairs.

"Really, Jeejee," said Mother crossly, "this has got to stop. There's not enough *of* you to fast."

Putting him to bed, Mother concocted huge strength-giving curries, only to have Jeejee complain that there was no Bombay duck, the dried fish which was such a pungent and attractive addition to any curry.

"But you can't get it here, Jeejee; I've tried," Mother protested.

Jeejee waved his hands like pale bronze moths against the white of the sheet.

"Fakyo tells that in life there is a substitute for everything," he said firmly.

So when he recovered sufficiently, he paid a visit to the fish market in the town and purchased a vast quantity of fresh sardines. We came back from a pleasant morning's shopping in the town to find the kitchen and its environs untenable. Jeejee, brandishing a knife with which he was gutting the fish before laying them out in the sun to dry outside the back door, was doing battle with what appeared to be every fly, bluebottle and wasp in the Ionian Islands. He had been stung about five times, and one eye was swollen and partially closed. The smell of rapidly decomposing sardines was overwhelming, and the kitchen floor and table were covered in snowdrifts of the silver fish skin and bits of entrails. It was only when Mother showed him the article on Bombay duck in the *Encyclopædia Britannica* that he reluctantly gave up the idea of sardines as a substitute. It took Mother two days, with buckets of hot water and disinfectant, to rid the kitchen of the smell, and even then there was still the odd wasp blundering in hopefully through the windows.

"Perhaps I'd better find you a substitute in Athens or Istanbul," said Jeejee hopefully. "I vas thinking that lobster baked and crushed to a powder. . . ."

"I wouldn't worry about it, Jeejee dear," said Mother hurriedly. "We've done without it for some time now, and it hasn't hurt us."

Jeejee was en route for Persia via Turkey in order to visit an Indian fakir practicing there.

"From him I shall learn many things to add to Fakyo," said Jeejee. "He is a great man. In particular, he is a great exponent of holding his breath and going into a trance. He vas vunce buried for a hundred and twenty days."

"Extraordinary," said Mother, deeply interested.

"You mean buried alive?" asked Margo. "Buried alive for a hundred and twenty days? How horrible! It doesn't seem natural somehow."

"But he's in a trance, dear Margo; he feels nothing," explained Jeejee.

"I'm not so sure," said Mother musingly. "That's why I want to be cremated, you know. Just in case I happen to slip into a trance and no one notices it."

"Don't be ridiculous, Mother," said Leslie.

"It's not ridiculous," said Mother firmly. "People are so careless nowadays."

"And what else does a fakir do?" asked Margo. "Can he make mango trees grow from seeds? You know, straight away? I saw them do that in Simla once."

"That is simple conjuring," said Jeejee. "What Andrawathi does is much more complex. He is an expert in levitation, for example, and it is vun of the things I vant to see him about."

"But I thought levitation was card tricks," said Margo.

"No," said Leslie, "it's floating about, sort of flying, isn't it, Jeejee?"

"Yes," said Jeejee. "A vonderful ability. A lot of the early Christian saints could do it. I myself have not yet reached that stage of proficiency; that is vhy I vant to study under Andrawathi."

"How lovely to be able to float about like a bird," said Margo delightedly. "What fun you could have."

"I believe it to be a truly tremendous experience," said Jeejee, his eyes shining. "You feel as if you are being lifted tovards heaven."

The following day, just before lunch, Margo came rushing into the drawing room in a state of panic.

"Come quickly! Come quickly!" she screamed. "Jeejee's committing suicide!"

We hurried outside and there, perched on the windowsill of his room, was Jeejee, clad in nothing but a loincloth.

"He's got one of those trances again," said Margo, as if it were an infectious disease.

Mother straightened her glasses and stared upwards. Jeejee started to sway gently.

"Go upstairs and grab him, Les," said Mother, "quickly. I'll keep him talking."

The fact that Jeejee was raptly silent did not occur to her. Leslie rushed into the house. Mother cleared her throat.

"Jeejee, dear," she fluted, "I don't think it's very wise of you to be up there. Why don't you come down and have lunch?"

Jeejee did come down, but not quite as Mother intended.

He stepped gaily out into space and, accompanied by horrified cries from Mother and Margo, fell earthwards. He crashed into the grapevine some ten feet beneath his window, sending a shower of grapes onto the flagstones. Fortunately the vine was an old and sinewy one and it held Jeejee's slight weight.

"My God!" he shouted. "Vere am I?"

"In the grapevine," screamed Margo excitedly. "You agitated yourself there."

"Don't move till we get a ladder," said Mother faintly.

We got a ladder and extricated the tousled Jeejee from the depths of the vine. He was bruised and scratched but otherwise unhurt. Everyone's nerves were soothed with brandy and we sat down to a late lunch. By the time evening came, Jeejee had convinced himself that he had in fact succeeded in levitating himself.

"If my toes had not become entangled in the pernicious vine, I vould have gone sailing around the house," he said, lying bandaged but happy on the sofa. "Vat an achievement!"

"Yes, well, I'll be happier if you don't practice while you are staying here," said Mother. "My nerves won't stand it."

"I vill come back from Persia and spend my birthday with you, my dear Mrs. Durrell," said Jeejee, "and I vill then report progress."

"Well, I don't want a repetition of today," said Mother severely. "You might have killed yourself."

Two days later Jeejeebuoy, still covered with sticking plaster but undaunted, left for Persia.

"I wonder if he will come back for his birthday," said Margo. "If he does, let's have a special party for him."

"Yes, that's a good idea," said Mother. "He's such a sweet boy, but so . . . erratic, so . . . *unsafe*."

"Well, he's the only guest we've had who could really be described as having paid a flying visit," said Leslie.

The Royal Occasion

Kings and bears oft worry their keepers.
SCOTTISH PROVERB

IN THOSE HALCYON DAYS we spent in Corfu, it could easily be said that every day was a special day, specially colored, specially arranged, and that it differed completely from the other three hundrd and sixty-four and was specially memorable because of this. But there is one day in particular which stands out in my mind, for it involved not only the family and their circle of acquaintances but the entire population of Corfu.

It was the day that King George returned to Greece, and nothing like it for color, excitement and intrigue had ever been experienced on the island. Even the difficulties of organizing St. Spiridion's procession paled into insignificance beside this event.

I first heard about the honor that was about to fall on Corfu from my tutor, Mr. Kralefsky, who was so overwhelmed with excitement that he took scant interest in the cock linnet I had been at considerable pains to procure for him.

"Great news, dear boy, great news! Good morning, good morning," he greeted me, his large soulful eyes brimming with tears of emotion, his shapely hands flapping to and fro and his head bobbing with excitement below his humpback. "A proud day for this island, by

Jove! Yes, indeed, a proud day for Greece, but an especially proud one for this, our island. Er . . . what? Oh, the linnet . . . yes. Nice birdie . . . tweet, tweet. But, as I was saying, what a triumph for us here in this little realm set in a sea of blue, as Shakespeare has it, to have the king visit us."

This, I thought, was more like it. I could raise a faint enthusiasm for a real king, if only for the fringe benefits that might accrue. Which king was it, I inquired, and would I have a holiday when he came?

"Why, the king of Greece, King George," said Mr. Kralefsky, rather shocked by my ignorance. "Didn't you know?"

I pointed out that we did not have the dubious benefits of a wireless in the country and so, for the most part, we lived in a state of blissful ignorance.

"Well," said Mr. Kralefsky, gazing at me rather worriedly, as if blaming himself for my lack of knowledge, "well, we had Metaxas, as you know, and he was a dictator. Now, mercifully, they've got rid of him, odious man, so now His Majesty can come back."

When, I inquired, had they got rid of Metaxas? Nobody had told me.

"Why, you remember, surely!" cried Kralefsky. "You must remember when we had the revolution and that cake shop was so badly damaged by the machine-gun bullets. Such unsafe things, I always think, machine guns."

Yes, I said, I did remember the revolution, because it had given me three days of blissful holiday from my lessons and the cake shop had been one of my favorite shops. But I had not connected this with Metaxas. Would there, I inquired hopefully, be another shop disemboweled by machine-gun fire when the king came?

"No, no," said Kralefsky, shocked. "No, it'll be a most

gay occasion. Everyone *en fête*, as they say. Well, it's such exciting news that I think we might be forgiven if we take the morning off to celebrate. Come upstairs and help me feed the birds."

So we made our way up to the huge attic room in which Kralefsky kept a vast collection of wild birds and canaries and spent a satisfying morning feeding them, Kralefsky dancing about the room waving the watering can, his feet scrunching on the fallen seed as if it were a shingle beach, singing snatches of the "Marseillaise" to himself.

Over lunch I imparted the news of the king's visit to the family, and they each received it in their characteristic ways.

"That'll be nice," said Mother. "I'd better start working out menus."

"He's not coming to stay *here*, thank God," Larry pointed out.

"I know that, dear," said Mother, "but . . . er . . . there'll be all sorts of parties and things, I suppose."

"I don't see why," said Larry.

"Because there always are," said Mother. "When we were in India, we always had parties during the durbar."

"This is not India," said Larry, "so I don't intend to waste my time working out the stabling for elephants. The whole thing will have a disruptive enough effect on the even tenor of our ways as it is, mark my words."

"If we're having parties, can I have some new clothes, Mother?" asked Margo eagerly. "I really haven't got a thing to wear."

"I wonder if they'll fire a salute," mused Leslie. "They've only got those old Venetian cannons, but I should think they'd be damned dangerous. I wonder if I ought to pop in and see the commandant of the fort."

"You keep out of it," Larry advised. "They want to welcome the man, not assassinate him."

"I saw some lovely red silk the other day," said Margo, "in that little shop . . . you know, the one where you turn right by Theodore's laboratory?"

"Yes, dear, how nice," said Mother, not listening. "I wonder if Spiro can get me some turkeys?"

But the effect of the royal visit on the family paled into insignificance in comparison with the traumatic effect it had on Corfu as a whole. It was pointed out, by somebody who should have known better, that not only was the island going to be graced by a visit from the monarch but the whole episode was particularly symbolic, as when the king arrived in Corfu he would be setting foot on Greek soil for the first time since his exile. At this thought the Corfiotes lashed themselves into a fever of activity, and before long so complicated, so acrimonious, and so incredible had the preparations become that we were forced to go into town each day to sit on the Platia with the rest of Corfu to learn the news of the latest scandal.

The Platia, laid out with its great arches to resemble the Rue de Rivoli by French architects in the early days of the French occupation of Corfu, was the hub of the island. Here you would sit at little tables under the arches or beneath the shimmering trees, and sooner or later you would see nearly everyone on the island and hear every facet of every scandal. One just simply sat there drinking quietly and, sooner or later, all the protagonists in the drama were washed up at one's table.

"I *am* Corfu," said Countess Malinopoulos. "Therefore it is incumbent upon me to form the committee that works out how we are to welcome our gracious king."

"Yes, indeed, I do see that," Mother agreed nervously.

The countess, who resembled a raddled black crow wearing an orange wig, was a formidable force, there was no doubt, but the matter was too important to allow her to ride rough-shod over everyone, and so within a very short time there were no less than six welcoming committees, all struggling to persuade the local nomarch that their plans ought to take preference over all others. It was rumored that he had an armed guard and slept in a locked room after an attempt by one of the female committee members to sacrifice her virginity in order to get his approval of her committee's schemes.

"Disgusting!" trumpeted Lena Mavrokondas, rolling her black eyes and smacking her red lips, wishing that she had thought of the idea herself. "Imagine, my dears, a voman of her age trying to break into the nomarch's room, naked!"

"It does seem a curious way to try to get his ear," Larry agreed innocently.

"No, no, it is too absurd," Lena went on, deftly popping olives into her scarlet mouth as though she were loading a gun. "I've seen the nomarch and I am sure he will agree to my committee being the official one. It is such shame the British flit is not in port; we could then have arrange a guard of honor. Oh, those lovely sailors in their uniform, they always look so clean and so virulent."

"The incidence of infectious diseases in the Royal Navy – " Larry began, when Mother hastily interrupted.

"Do tell us what your plans are, Lena," she said, glaring at Larry, who was on his eighth ouzo and inclined to be somewhat unreliable.

"Soch plans, my dears, soch plans ve 'ave! This whole Platia vill be decorate in blue and vite, but alvays ve 'ave troubles with that fool Marko Panioti," said Lena, rolling her eyes in despair.

Marko, we knew, was a sort of inspired madman, and we wondered how he had got onto the committee at all.

"What does Marko want to do?" asked Larry.

"Donkeys!" hissed Lena, as if it were an obscene word.

"Donkeys?" repeated Larry. "He wants to have donkeys? What does he think it is? An agricultural show?"

"This I explain 'im," said Lena, "but alvays 'e vants to 'ave donkeys. 'E says it is symbolic, like Christ's ride into Jerusalem, so 'e vants blue and vite donkeys."

"Blue and white ones? You mean dyed?" asked Mother. "Whatever for?"

"To match the Greek flags," said Lena, rising to her feet and facing us grimly, shoulders back, hands clenched. "But I tell 'im, 'Marko,' I say, 'you 'ave donkeys over my dead corpse.'"

She strode off down the Platia, every inch a daughter of Greece.

The next one to stop at our table was Colonel Velvit, a tall, rather beautiful old man with a Byronic profile and an angular body that twitched and moved like a windblown marionette. With his curling white hair and his flashing dark eyes, he looked rather incongruous in his Scout uniform, but he carried it off with dignity. Since his retirement, his one interest in life was the local Scout troop, and while there were those unkind enough to say that his interest in Scouts was not entirely altruistic, it could be said he worked hard and had never been caught.

He accepted an ouzo and sat mopping his face with a lavender-scented handkerchief.

"Those boys," he said plaintively, "those boys of mine will be the death of me. They are so high-spirited."

"What they probably need is a bevy of nubile Girl Guides," said Larry. "Have you thought of that?"

"It is no joke, my dear," said the colonel, eyeing Larry morosely. "They are so full of high spirits I fear they will get up to some prank or other. I was simply horrified at what they did today, and the nomarch was most annoyed."

"The poor nomarch appears to be getting it in the neck from every direction," said Leslie.

"What did your Scouts do?" asked Mother.

"Well, as you know, my dear Mrs. Durrell, I am training them to put on a special demonstration for His Majesty on the evening of his arrival," said the colonel, sipping his drink delicately like a cat. "First they march out, some dressed in blue and some in white, in front of the . . . how do you call it? . . . dais! Exactly so, the dais. And they form a square and salute the king. Then, at the word of command, they change positions and form the Greek flag. It's a very striking sight, though I say it myself."

He paused, drained his glass and sat back.

"Well, the nomarch wanted to see how we were progressing, so he came along and stood on the dais, representing the king, as it were. Then I gave the command and the troop marched out."

He closed his eyes and a small shudder shook him.

"Do you know what they did?" he asked in a small voice. "I have never felt so ashamed. They marched out, stopped in front of the nomarch and gave the Fascist salute. Boy Scouts! The Fascist salute!"

"Did they shout, 'Heil nomarch'?" asked Larry.

"Mercifully, no," said Colonel Velvit. "For a moment I was paralyzed with shock, and then, hoping that the nomarch had not noticed, I gave the command to form the flag. They moved about and, to my horror, the nomarch was confronted by a blue and white swastika.

The nomarch was furious. He almost canceled our part in the proceedings. What a blow to the Scout movement that would have been!"

"Yes, indeed," said Mother, "but they're only children, after all."

"That's true, my dear Mrs. Durrell, but I cannot have people saying that I am training a group of Fascists," said Colonel Velvit earnestly. "They'll be saying that I plan to take over Corfu."

During the ensuing days, as the time grew nearer for the great event, the island's inhabitants grew more and more frenzied and tempers grew shorter and shorter. Countess Malinopoulos was now no longer speaking to Lena Mavrokondas, who in turn was not speaking to Colonel Velvit because his Boy Scouts had given her a gesture of unmistakably biological nature as they passed by her house. All the leaders of the village bands, who always took part in the St. Spiridion's procession, had quarreled bitterly with each other over procedure in the march-past, and one evening on the Platia we were all treated to the sight of three incensed tuba players chasing a bass drummer, all in full uniform and carrying their instruments. The tuba players, obviously driven beyond endurance, cornered the drummer, tore his instrument from him and jumped on it. Immediately the Platia was a seething mass of infuriated bandsmen locked in combat. Mr. Kralefsky, who was an innocent bystander, received a nasty cut on the back of the head from a flying cymbal, and old Mrs. Kukudopoulos, who was exercising her two spaniels between the trees, had to pick up her skirts and run for it. This incident (everybody said when she died the next year) took years off her life, but as she was ninety-five when she died, this was scarcely credible. Soon nobody was on speaking terms with anybody, though they all talked to us, for we kept strictly neutral. Captain Creech, whom no one suspected of possessing a patriotic streak of any sort, was wildly excited by the whole thing and, to everyone's annoyance, went from committee to committee spreading gossip, singing bawdy songs, pinching unsuspecting and unprotected bosoms and buttocks, and generally making a nuisance of himself.

"Disgusting old creature!" said Mother, her eyes flashing. "I do wish he'd behave himself. After all, he is meant to be British."

"He's keeping the committees on their toes, if I may use the phrase," said Larry. "Lena tells me that her bottom was black-and-blue after the last meeting he attended."

"Filthy old brute," said Mother.

"Don't be so harsh, Mother," said Larry. "You know you're only jealous."

"Jealous!" squeaked Mother, bristling like a diminutive terrier. "Jealous! . . . of that . . . old . . . old . . . *libertine!* Don't be so disgusting. I won't have you say things like that, Larry, even in joke."

"But it's unrequited love for you that makes him drown his sorrows in wine and women," Larry pointed out. "If you'd make an honest man of him, he'd reform."

"He was drowning his sorrows in wine and women long before he met me," said Mother, "and as far as I'm concerned, he can go on doing so. He's one person I'm not interested in reforming."

The captain, however, was oblivious to all criticism.

"Darling girl!" he said to Mother the next time he met her. "You haven't by any chance a Union Jack in your bottom drawer?"

"No, Captain, I'm afraid I haven't," said Mother with dignity. "Neither have I a bottom drawer."

"What? A fine wench like you? No bottom drawer?

No nice collection of frilly black knickers to drive your next husband mad?" asked Captain Creech, eyeing Mother with a lecherous and rheumy eye.

Mother blushed and stiffened.

"I have no intention of driving *anyone* mad, with or without knickers!" she said with great dignity.

"That's my wench," said the captain, "game, that's what you are, game. I like a little nudity myself, to tell the truth."

"What d'you want a Union Jack for?" asked Mother, frigidly changing the subject.

"To wave, of course," said the captain. "All these wogs will be waving their flag, so we must show 'em the good old Empire's not to be overlooked."

"Have you tried the consul?" asked Mother.

"Him?" said the captain scornfully. "He said there was only one on the island and *that* was only to be used for special occasions. Whores' hair! I told him, if this is not a special occasion, what in the name of the testicles of St.Vitus is? So I told him to use his flagpole as an enema."

"I do wish you wouldn't encourage that dirty old man to come and sit with us, Larry," said Mother plaintively when the captain had staggered off in pursuit of the Union Jack. "His conversation is positively obscene, and I don't like him saying things like that in front of Gerry."

"It's your fault. You encourage him," said Larry. "All this talk about removing your knickers."

"Larry! You know perfectly well what I meant. It was a slip of the tongue."

"But it gave him hope," said Larry. "You'd better watch out or he'll be into your bottom drawer like a truffle hound, choosing nighties for the wedding night."

"Oh, do be quiet!" said Mother crossly. "Really, Larry, you do make me angry sometimes."

The island became more and more tense. From the remote mountain villages, where the older women were polishing up their cow horn headdresses and ironing their handkerchiefs, to the town, where every tree was being pruned and every table and chair on the Platia was being repainted, all was seething with acrimonious activity.

In the old part of the town, where the streets were two donkeys wide and the air was always redolent of freshly baked bread, fruit, sunshine and drains in equal quantities, was the tiny cafe belonging to a friend of mine, Costi Avgadrama. The cafe was justly famous for producing the best ice cream in Corfu, for Costi had been to Italy and had learned all the dark arts of ice cream making. His confections were much in demand, and there was scarcely a party worth calling a party given on the island that did not include one of Costi's enormous, tottering, multicolored creations. Costi and I had a good working agree-ment; I would go to his cafe three times a week to collect all the cockroaches in his kitchen to feed my birds and animals, and in return for this service I was allowed to eat as many ice creams as I could during my work. Determined that his shop should be clean for the royal visit, I went along to Costi's cafe about three days before the king was due and found him in a mood of suicidal despair such as only a Greek, with the aid of ouzo, can acquire and sustain. I asked him what was the matter.

"I am ruined," he said sepulchrally, setting before me a stone bottle of ginger beer and a gleaming white ice cream big enough to sink the *Titanic*. "I am a ruined man, *kyria* Gerry. I am a laughingstock. No longer will people say, 'Ah, Corfu, that is where Costi's ice cream comes from.' No, they will say instead, 'Corfu? That's where that fool Costi's ice cream comes from.' I shall have to leave the island, there is no other course. I shall go to Zante or maybe Athens, or perhaps I shall join a monastery. My wife and children will starve, my poor old parents will feel burning shame as they beg for their bread – "

Interrupting these gloomy prophecies, I asked what had happened to bring about this state of despair.

"I am a genius," said Costi simply and without boastfulness, seating himself at my table and absentmindedly pouring himself out another ouzo. "No one in Corfu could produce ice creams like mine, so succulent, so beautiful, so . . . so *cold*."

I said this was true. I went further, for he obviously needed encouragement; I said that his ice creams were famous throughout Greece, maybe even throughout Europe.

"True," groaned Costi, "so it was natural that when the king was to visit Corfu, the nomarch wanted him to taste my ice cream."

I was greatly impressed and said so.

"Yes," said Costi, "twelve kilos of ice cream I was to deliver to the palace at *Mon Repos* and one special ice cream for the great banquet on the night His Majesty arrives. Aghh! it was this special ice that was my undoing. This is why my wife and children must starve. Ah, cruel and relentless fate!"

Why? I asked bluntly, through a mouthful of ice cream. I was in no mood for the frills; I wanted to get to the core of the story.

"I decided that this ice cream must be something new, something unique, something never dreamt of before," said Costi, draining his ouzo. "All night I lay awake waiting for a sign."

He closed his eyes and turned his head from side to side on a hot, unyielding imaginary pillow.

"I did not sleep, I was in a fever. Then, just as the

first cocks crowed, *'Ku-ka-ra-ka, kooo,'* I was blinded by a flash of inspiration."

He smote himself so hard on the forehead he almost fell out of his chair. Shakily, he poured out another ouzo.

"I saw before my hot and tired eyes the vision of a flag, a flag of Greece, the flag for which we have all suffered and died, but the flag made in my *best superior quality, full cream ice cream*," he said triumphantly, and sat back to see its effect on me.

I said I thought the idea was the most brilliant I had ever heard of. Costi beamed, and then, remembering, his expression became one of despair again.

"I leapt out of bed," he continued dolefully, "and ran into my kitchen. There I discovered that I had not the ingredients to carry out my plan. I had chocolate to color the cream brown, I had dyes to make it red or green or even yellow, but I had nothing, nothing at all, to make the blue stripes in the flag." He paused, drank deeply, and then drew himself up proudly.

"A lesser man . . . a . . . Turk or an Albanian . . . would have abandoned the plan. But not Costi Avgadrama. You know what I did?'

I shook my head and took a swig of ginger beer.

"I went to see my cousin Michaeli, you know, he works for the chemist's down by the docks. Well, Michaeli – may St. Spiridion's curse fall upon him and his offspring – gave me some stuff to make the stripes blue. Look!" said Costi.

He went to his cold room and disappeared inside; then he came staggering out bearing a mammoth dish and laid it in front of me. It was full of ice cream in blue and white stripes and did look remarkably like the Greek flag, even if the blue was a little on the purple side. I said I thought it was magnificent.

"Deadly!" hissed Costi. "Deadly as a bomb."

He sat down and stared malevolently at the huge dish. I could see nothing wrong with it except that the blue was more a methylated spirits sort of color than true blue.

"Disgraced! By my own cousin, by that son of an unmarried father!" said Costi. "He gave me the powder, he said it would be fine; he promised me, the viper tongue, that it would work."

But it had worked, I pointed out, so what was the trouble?

"By God and St. Spiridion's mercy," said Costi piously, "I had the idea of making a small flag for my family, just so they could celebrate their father's triumph. I cannot bear to think what would have happened if I had not done this."

He rose to his feet and opened the door leading from the cafe to his private quarters.

"I will show you what that monster, my cousin, has done," he said, and called up the stairs, "Katarina! Petra! Spiro! Come!"

Costi's wife and his two sons came slowly and reluctantly down the stairs and stood in front of me. To my astonishment I saw that they all had bright purple mouths, the rich, royal purple of a summer beetle's wing case.

"Put out your tongue," Costi commanded.

The family opened their mouths and poked out tongues the color of a Roman's robe. They looked like rather macabre sorts of orchids, or a species of mandrake, perhaps. I could see Costi's problem. In the helpful, unthinking way that Corfiotes have, his cousin had given him a packet of gentian violet. I had once had to paint a sore on my leg with this substance so I knew that, among its many properties, it was an extremely tenacious dye. Costi would have a purple wife and children for some weeks to come.

"Just imagine," he said to me in a hushed whisper, having sent his discolored wife and brood back upstairs, "just imagine if I'd sent this to the palace. Imagine all those church dignitaries, their beards purple! A purple nomarch and a purple king! I would have been shot."

I said I thought it would have been rather funny. Costi was greatly shocked. When I grew up, he said severely, I would realize that some things in life were very serious, not comical.

"Imagine the reputation of the *island* . . . imagine *my* reputation if I had turned the king purple," he said, as he gave me another ice cream to show that there was no ill-feeling. "Imagine how the foreigners would have laughed if the Greek king turned purple. Po! po! po! po! St. Spiridion save us!"

And how about the cousin, I inquired; how had he taken the news?

"He doesn't know yet," said Costi, grinning evilly, "but he will soon. I've just sent him an ice cream shaped like the Greek flag."

So the island was wound up to a pitch of unbearable excitement and irritation when the great day dawned. Spiro had arranged his huge, ancient Dodge with the hood down as a sort of combination grandstand and battering ram, determined that the family, at any rate, were going to get a good view of the proceedings. In a festive mood, we drove into town and had a drink on the Platia to pick up the news of the final preparations. Lena, resplendent in green and purple, told us that Marko had finally, if reluctantly, given up his idea of blue and white donkeys but now had another plan only slightly less bizarre.

"You know 'e 'as 'is father's printing works, huh?"

said Lena. "Vell, 'e say 'e is to print thousands and thousands of Greek flags and take them out in 'is yacht and then scatter them over the vater so that the king's ship 'as a carpet of Greek flags to sail on, no?"

Marko's yacht was the joke of Corfu; a once rather lush cabin cruiser, Marko had added so much superstructure to it that, as Leslie rightly said, it looked like a sort of seagoing Crystal Palace with a heavy list to starboard. Every time Marko set sail in it, bets were laid as to when and if he would return.

"So," continued Lena, "first 'e 'ave the flags print, then 'e finds they don't float – they sink. So 'e makes little crosses of vood and sticks the flags on them so that they vill float."

"It sounds rather a nice idea," said Mother.

"If it works," said Larry. "You know Marko's genius for organization. Remember Constantine's birthday."

In the summer Marko had organized a sumptuous picnic for his nephew Constantine's birthday. It would have been a splendid event, with everything from roast suckling pig to watermelons filled with champagne. The elite of Corfu were invited. The only snag was that Marko had got his beaches muddled, and while he sat in solitary splendor surrounded by enough food to feed an army on a beach far down south, the elite of Corfu, hot and hungry, waited on a beach in the far north of the island.

"Vell," said Lena, with an expressive shrug, "ve cannot stop him. All the flags are loaded on his boat. He has sent a man with a rocket to Coloura."

"A man with a rocket?" asked Leslie. "What for?"

Lena rolled her eyes expressively.

"When the man sees the king's ship, he fires the rocket," she said. "Marko sees the rocket, and this gives him time to rush out and cover the sea with flags."

"Well, I hope he succeeds," said Margo. "I like Marko."

"My dear, so do ve all," said Lena. "In my village vhere I 'ave my villa ve have a village idiot. He is a charming, *très sympathique*, but ve do not vant to make him the mayor."

With this waspish parting shot, she left us. The next one to arrive was Colonel Velvit in an agitated state.

"You haven't by any chance seen three small, fat Boy Scouts?" he asked. "No, I didn't think you would have. Little brutes! They went off into the country *in their uniforms*, the little savages, and came back looking like pigs! I sent them off to the cleaners to get their uniforms cleaned and they've disappeared."

"If I see them, I'll send them to you," said Mother soothingly. "Don't worry."

"Thank you, my dear Mrs. Durrell. I would not worry, but the little devils are an important part of the proceedings," said Colonel Velvit, preparing to go in search of the missing Scouts. "You see, not only do they form part of the stripe in the flag but they have to demolish the bridge as well."

With this mysterious remark, he departed, loping off like a hound.

"Bridge? What bridge?" asked Mother, bewildered.

"Oh, it's part of the show," said Leslie. "Among other things, they build a pontoon bridge over an imaginary river, cross it, and then blow it up to prevent the enemy following."

"I always thought Boy Scouts were peaceful," said Mother.

"Not the Corfiote ones," said Leslie. "They're probably the most warlike inhabitants of Corfu."

At that moment Theodore and Kralefsky, who were to share the car with us, arrived.

"There has been . . . er . . . you know . . . a slight hiatus over the salute," Theodore reported to Leslie.

"I knew it!" said Leslie angrily. "That fool of a commandant! He was too airy-fairy when I spoke to him. I told him those Venetian cannons would burst."

"No, no . . . er . . . the cannon haven't burst. Er . . . um . . . at least, not *yet*," said Theodore. "No, it is a problem of timing. The commandant was very insistent that the salute should be fired the moment the king's foot touches Greek soil. The . . . er . . . um . . . difficulty was apparently to arrange a signal from the docks that could be seen by the gunners in the . . . er . . . you know . . . the fort."

"So what have they arranged?" asked Leslie.

"They have sent a corporal down to the docks with a forty-five," said Theodore. "He is to fire it at the moment *before* the king sets foot on the shore."

"Does he know how to fire it?" asked Leslie sceptically.

"Well . . . er," said Theodore, "I had to spend quite some time trying to make him see that it was dangerous to put it . . . um . . . you know . . . loaded *and* cocked into his holster."

"Silly fool, he'll shoot himself through the foot like that," said Leslie.

"Never mind," said Larry, "there's bound to be some bloodletting before the day is out. I hope you brought your first-aid kit, Theodore."

"Don't say things like that, Larry," said Mother. "You make me feel quite nervous."

"Ifs you're readys, Mrs. Durrells, we oughts to gets going," said Spiro, who had appeared, brown, scowling, looking like a gargoyle on holiday from Notre Dame. "The crowd's getting very tense."

"Dense, Spiro, *dense*," said Margo.

"That's what I says, Misses Margo," said Spiro. "But don'ts you worry. I'll fix 'em. I'll scarce them out of the way with my horn."

"Spiro really ought to write a dictionary," said Larry as we climbed into the Dodge and wedged ourselves onto the capacious leather seats.

Since early morning the white dusty roads had been jammed with carts and donkeys bringing peasants into the capital for the great event, and a great pall of dust covered the countryside, turning the plants and trees by the roadside white, hanging in the air like microscopic snow. The town was now as full or fuller than it was on St. Spiridion's Day, and great bevies of people were eddying across the Platia in their best clothes like clouds of windswept blossoms. Every back street was jammed with humanity mixed with donkeys, the whole moving at a glacier pace, and the air was full of excited chatter and laughter, the pungent smell of garlic, and the all-pervading smell of mothballs, the sign of special clothes carefully extracted from their places of safekeeping. On every side you could hear brass bands tuning up, donkeys braying, the cries of the street vendors, and the excited screams of children. The town quivered and throbbed like a great, multicolored, redolent beehive.

Driving at a snail's pace, honking his great, rubber-bulbed horn to "scarce" the uncaring populace out of the way, Spiro drove us down to the docks. Here all was bustle and what passed for efficiency; a band was lined up, its instruments sparkling, its uniforms immaculate, its air of respectability only slightly marred by the fact that two of its members had black eyes. Next to it was a battalion of local soldiery, looking remarkably clean and neat. Church dignitaries, with their carefully combed white, silver and iron-grey beards, bright and gay as a flock of parrots in their robes, chatted animatedly to each other, stomachs bulging, beards wagging, plump, well-manicured hands moving in the most delicate of gestures. Near the dockside where the king would come ashore stood a forlorn-looking corporal; obviously his responsibilities were weighing heavily on him for he kept fingering his revolver holster nervously and biting his nails.

Presently there was a surge of excitement, and everyone was saying, "The king! The king! The king is coming!" The corporal adjusted his hat and stood a little straighter. What had given rise to this rumor was the sight of Marko Panioti's yacht putting out into the bay and lumbering to and fro while Marko, in the stern, could be seen unloading bundle after bundle of Greek flags.

"I didn't see the rocket, did you?" asked Margo.

"No, but you can't see the headland from here," said Leslie.

"Well, I think Marko's doing it splendidly," said Margo.

"It's certainly a very pretty effect," said Mother.

And indeed it was, for several acres of the smooth sea were covered with a carpet of tiny flags, which looked most impressive. Unfortunately, as we were to learn within the next hour and a half, Marko's timing had been faulty. The man he had stationed up in the north of the island to fire the signal rocket was most reliable, but his identification of ships left a lot to be desired, and so what eventually appeared was not the ship conveying the king but a rather grubby little tanker on its way to Athens. This in itself would not have been such a grave error, but Marko, carried away as so many Corfiotes were that day, had failed to check on the glue with which the flags were stuck to the little wooden pieces that allowed them to float. So as we waited for the king, we were treated to the sight of the glue disintegrating under the influence of seawater and several thousand Greek flags sinking ignominiously to the bottom of the bay.

"Oh, poor Marko, I feel so sorry for him," said Margo, almost in tears.

"Never mind," said Larry consolingly, "perhaps the king likes little bits of wood."

"Um . . . I don't . . . you know . . . think so," said Theodore. "You see how they're all shaped like a little cross. That in Greece is considered a very bad omen."

"Oh, dear," said Mother, "I do hope the king won't realize that Marko did it."

"If Marko's wise, he'll go into voluntary exile," said Larry.

"Ah, here he comes at last," said Leslie as the king's ship sailed majestically across several acres of little wooden crosses, as though ploughing its way through a marine war cemetery, and tied up.

The gangplank was lowered, the band struck up blaringly, the army came to attention, and the crowd of church dignitaries moved forward like a suddenly uprooted flower bed. They reached the bottom of the gangplank, the band stopped playing, and to a chorus of delighted Ah's the king made his appearance, paused briefly to salute, and then made his way slowly down the gangway. It was the little corporal's great moment. Sweating profusely, he had moved as close to the gangplank as he could and he had his gaze riveted on the king's feet. His instructions had been explicit; three paces before the king stepped off the gangway and onto Greek soil, he was to give the signal. This would give the fort enough time to fire the cannon as the king stepped ashore.

The king descended slowly. The atmosphere was

tense with emotion. The corporal fumbled with his holster and then, at the crucial moment, drew his forty-five and fired five rounds off approximately two yards away from the king's right ear. It immediately became obvious that the fort had not thought to tell the welcoming committee about its signal, so the committee, to say the least, was taken aback, as was the king and, indeed, as we all were.

"My God, they've amputated him," screamed Margo, who always lost both her head and her command over English in moments of crisis.

"Don't be a fool, it's the signal," snapped Leslie, training his field glasses on the fort.

But it was obvious that the welcoming committee thought much the same as my sister. As one man, they fell on the unfortunate corporal. He, white-faced and protesting, was pummeled and thumped and kicked, his revolver was torn from his grasp, and he was hit smartly on the head with it. It is quite probable that he would have come to some serious harm if at that moment the cannons had not blared out in an impressive cumulous cloud of smoke from the ramparts of the fort and vindicated his action.

After this all was smiles and laughter, for the Corfiotes had a keen sense of humor; only the king looked a trifle pensive. He climbed into the official open car and a snag made its appearance; for some reason the door would not lock. The chauffeur slammed it, the sergeant in charge of the troops slammed it, the band leader slammed it, and a passing priest slammed it, but it refused to stay shut. The chauffeur, not to be defeated, backed up and took a run and kicked it violently. The car shuddered, but the door remained obdurate. They tried string, but there was nothing to tie it to. Eventually, since there could be no further delay, they were forced to drive off with the nomarch's secretary hanging over the back of the seat and holding the door shut with one hand.

Their first stop was at St. Spiridion's Church so that the king could make his obeisances to the mummified remains of the saint. Surrounded by a forest of ecclesiastical beards, he disappeared into the dark depths of the church, where a thousand candles bloomed like a riot of primroses. It was a hot day, and the chauffeur of the king's car was feeling a bit exhausted after his fight with the door, so without telling anyone, he left the car parked in front of the church and nipped round the corner for a drink. And who is to blame him? Who, on occasions such as this, has not felt the same? However, his estimation of the time the king would take to visit the saint was inaccurate, so when the king, surrounded by the cream of the Greek Church, suddenly emerged from the church and took his place in the car, the chauffeur was conspicuous by his absence. As was usual in Corfu when a crisis was reached, everyone blamed everyone else for the chauffeur's disappearance. A quarter of an hour passed while accusations were hurled, fists were shaken, and runners were sent in all directions in search of the chauffeur.

There was some delay because no one knew which cafe he was honoring with his presence, but eventually he was tracked down, a stream of vituperation was poured on his head, and he was dragged ignominiously away in the middle of his second ouzo.

The next stop was the Platia, where the king was to see the march-past of troops and bands and the exhibition by the Scouts. By driving cacophonously through the narrow back streets, Spiro got us to the Platia long before the king's car.

"Surely they can't do anything else wrong," said Mother worriedly.

"The island has surpassed itself," said Larry. "I had hoped they might get a puncture between the docks and the church, but that was asking too much, I suppose."

"Well, I'm not so sure," said Theodore, his eyes twinkling. "Remember, this is Corfu. They might well have something more in store for us."

"I do hope not," said Kralefsky. "Really! Such organization! It makes one blush."

"They can't think up anything more, Theo, surely," Larry protested.

"I wouldn't like to be . . . er . . . um . . . bank on it . . . you know," said Theodore.

As it turned out, he was perfectly right.

The king arrived and took his place on the dais. The troops marched past with great vigor and all of them managed to be more or less in step. Corfu was rather a remote garrison in those days and the recruits did not get much practice, but nevertheless they acquitted themselves very creditably. Next came the mass bands – bands from every village on the island, their variously colored uniforms glowing, their instruments so polished the gleam of them hurt the eyes. If their delivery quavered a little and was slightly off-key, it was more than made up for by the volume and force of their playing.

Then it was the turn of the Scouts, and we all clapped and cheered as Colonel Velvit, looking like an extremely nervous and attenuated Old Testament prophet in Scout's uniform, led his diminutive troops onto the dusty Platia. They saluted the king and then, obeying a rather strangled falsetto order from the colonel, they shuffled to and fro and formed the Greek flag. Such a wave of clapping and cheering broke out that it must surely have been heard in the remotest vastnesses of the Albanian mountains. After a short display of gymnastics, the troops went over to where two white lines represented the two banks of a river. Here half the troop hurried away and reappeared with planks necessary for making a pontoon bridge while the other half were busy getting a line across the treacherous waters. So fascinated were the crowd by the mechanics of this that they drifted closer and closer to the river, accompanied by the policemen who were supposed to be keeping them back. In record time, the Scouts, none of whom were more than eight years old, created their pontoon across the imaginary river. Then, led by one small boy blowing vociferously and inaccurately on a trumpet, they jog-trotted across the bridge and stood at attention on the opposite side. The crowd were enchanted; they clapped, cheered, whistled and stamped. Colonel Velvit allowed himself a small, tight, military smile and cast a proud look in our direction. Then he barked out a word of command. Three small, fat Scouts detached themselves from the troop and made their way to the bridge carrying fuses, a plunger and the other demolition equipment. They fixed everything up and then rejoined the troop, unwinding the fuse wire as they went. They stood at attention and waited. Colonel Velvit savored his big moment; he glanced round to make sure he had everyone's undivided attention. The silence was complete.

"Demolish bridge!" roared Colonel Velvit, and one of the Scouts crouched and pressed the plunger home.

The next few minutes were confused, to say the least.

There was a colossal explosion; a cloud of dust, gravel and bits of bridge was thrown into the air, to descend like hail upon the population. The first three rows of the crowd, all the policemen and Colonel Velvit were thrown flat on their backs. The blast, carrying with it gravel and splinters of wood, reached the car where we were sitting, battered against the coachwork like machine-gun fire and blew Mother's hat off.

"Christ Almighty!" said Larry. "What the hell's that fool Velvit playing at?"

"My hat," panted Mother. "Somebody get my hat!"

"I'll gets it, Mrs. Durrells, don'ts you worrys," roared Spiro.

"Most unnerving, most unnerving," said Kralefsky, his eyes closed, mopping his brow with his handkerchief. "Far too militant for small boys."

"Small boys! Bloody little fiends," said Larry angrily, shaking gravel out of his hair.

"I felt sure something else would happen," said Theodore with satisfaction, happy now that Corfu's reputation for calamities was secure.

"They must have had some sort of explosives," said Leslie. "I can't think what Colonel Velvit was playing at. Damned dangerous."

It became obvious a little later that it was not the colonel's fault. Having rather shakily lined up his troop and marched them away, he returned to the scene of the carnage to apologize to Mother.

"I cannot tell you how mortified I am, Mrs. Durrell," he said, tears in his eyes. "Those little brutes got some dynamite from some fishermen. I assure you, I knew absolutely nothing about it, nothing."

In his dust-stained uniform and his battered hat, he looked very pathetic.

"Oh, don't worry, Colonel," said Mother, shakily lifting a brandy and soda to her lips, "it's the sort of thing that could happen to anyone."

"Happens all the time in England," said Larry. "Never a day passes – "

"Do come and have dinner with us," interrupted Mother, giving Larry a quelling look.

"Thank you, dear lady, you are too kind," said the colonel. "I must go and change."

"I was very interested in the reaction of the spectators," said Theodore with scientific relish. "You know . . . er . . . the ones who were blown down."

"I should think they were damned annoyed," said Leslie.

"No," said Theodore proudly, "this is Corfu. They all . . . you know . . . helped each other up, brushed each other down, and remarked on how good the whole thing was . . . er . . . how realistic. It didn't seem to occur to them that there was anything strange in Boy Scouts having dynamite."

"Well, if you live long enough in Corfu, you cease to be surprised at anything," said Mother with conviction.

Eventually, after a prolonged and delicious meal in town, during which we tried to convince Colonel Velvit that his bridge demolition had been the high spot of the day, Spiro drove us home through the cool, velvety night. The scops owls called *toink toink* to each other, chiming like strange bells among the trees; the white dust billowed up behind the car and remained suspended like a summer's cloud in the still air; the dark cathedral groves of the olives were pricked out with the pulsing green lights of fireflies. It had been a good, if exhausting, day and we were glad to be home.

"Well," said Mother, stifling a yawn as she picked up her lamp and made her way to the stairs, "king or no king, I'm staying in bed until twelve tomorrow."

"Oh," said Larry contritely, "didn't I tell you?"

Mother paused halfway up the stairs and looked at Larry, the wavering lamplight making her shadow quiver and leap on the white wall.

"Tell me what?" she asked suspiciously.

"About the king," said Larry. "I'm sorry. I should have told you before."

"Told me what?" said Mother, now seriously alarmed.

"I've asked him to lunch," said Larry.

"Larry! You haven't! Really, you are thoughtless. . . ." Mother began, and then realized that she was having her leg pulled.

She drew herself up to her full five feet.

"I don't think that's funny," she said frigidly, "and anyway, the laugh would have been on him, for I've only got eggs in the house."

With great dignity, ignoring our laughter, she made her way to bed.

The Paths of Love

Stay me with flagons, comfort me with apples: for I am sick of love.
THE SONG OF SOLOMON 2:5

IT HAD BEEN a prodigious, desiccating, earth-cracking summer that was so hot it even bleached the sky to a pale forget-me-not color and flattened the sea so that it lay like a great blue pool, unmoving, warm as fresh milk. At night you could hear the floors and shutters and beams of the villa shifting and groaning and cracking in the warm air as the last juices were sucked out of them. The full moon would rise like a red coal, glowering down at us from the hot, velvety sky, and in the morning the sun was already too warm to be comfortable ten minutes after it appeared. There was no wind, and the heat pressed down on the island like a lid. On the hillside in the breathless air, the plants and grasses withered and died. They stood bleached as blonde as honey, crisp as wood shavings. The days were so hot that even the cicadas started singing earlier and siestaed during the heat of the day, and the ground was so baked that there was nowhere you could walk without shoes.

The villa represented to the local animal life a series of large wooden caves which were perhaps half a degree cooler than the surrounding olive, orange and lemon groves, and so they flocked to join us. At first I was naturally blamed for this sudden influx of creatures, but eventually the invasion became so comprehensive that even my family realized I could not be responsible for quite such a large quantity and variety of life forms. Battalions of black ticks marched into the house and beset the dogs, massing in such numbers on their ears and heads it was like chain mail and just as difficult to remove. In desperation we had to douse them with kerosene, which made the ticks drop off so that we could deal with them. The dogs, deeply insulted by this treatment, slouched, panting, round the house, reeking of kerosene, dropping ticks in vast quantities. Larry suggested that we put up a notice saying "Danger – Inflammable Dogs," for, as he rightly pointed out, if anyone lit a match near one of them, the whole villa was liable to go up in flames like a tinderbox. But the kerosene only gave us a temporary respite. More and more ticks marched into the house until at night one could lie in bed and watch rows and rows of them on the walls performing strange route marches around the room. The ticks, fortunately, did not attack us but confined themselves to driving the dogs mad. However, the hordes of fleas that decided to take up residence with us were another matter. They arrived suddenly, out of nowhere, it seemed, like the Tartar hordes and overran us before we realized what was happening. They were everywhere. You could feel them hopping onto you and running up your legs as you walked around the house. The bedrooms became untenable, and for a time we took our beds out onto the broad verandahs and slept there.

But the fleas were not the most objectionable of the lesser inhabitants of the house. Tiny scorpions, black as ebony, infested the bathroom, where it was cool. Leslie, going in late one night to clean his teeth, was ill advised enough to go barefoot and was stung on the toe. The scorpion was only half an inch long, but the agony of the bite was out of all proportion to the size of the beast and it was some days before Leslie could walk. The larger scorpions preferred the kitchen area, where they would quite blatantly sit on the ceiling looking like misshapen aerial lobsters.

At night when the lamps were lit, thousands of insects appeared; moths of all shapes, from tiny fawn-colored ones with wings shaped like tattered feathers to the great big, striped, pink and silver hawk moths, whose death dives at the light were capable of breaking the lantern chimneys. Then there were the beetles, some as black as mourners, some gaily striped and patterned, some with short club-shaped antennae, others with antennae as long and thin as a mandarin's moustache. With these came a multitude of lesser forms of life, most of them so small that you needed a magnifying glass to make out their incredible shapes and colors.

Naturally, this conglomeration of insects was marvelous as far as I was concerned, and each evening I hung about the lights, my collecting boxes and bottles at the ready, vying with the other predators for choice specimens. And I had to look sharp, for the competition was brisk. On the ceiling were the geckoes, pale, pink-skinned, spread-fingered, bulbous-eyed, stalking the moths and beetles with minuscule care. Alongside them were the green, swaying, hypocritical mantises with their mad eyes and chinless faces, moving on slender, prickly legs like green vampires. On the ground level I had to contend with the enormous chocolate-colored spiders like lanky, furry wolves, who would lurk in the shadows and scuttle out and snatch a specimen almost from my very fingers. They were aided and abetted by the fat map toads in their handsome patchwork skins of green and silvery grey, who hopped and gulped their way, wide-eyed with astonishment, through this largesse of food, and the swift, furtive, and somehow sinister scutiger. This form of centipede has a body some three inches long, as thick as a pencil and flattened; around the perimeter is a hedge, a fringe of long, slender legs. When the creature moves, as each pair of legs comes into action, these fringes appear to undulate in waves as the animal progresses as smoothly as a stone on ice, silent and unnerving, for scutiggeria are among the more ferocious and skillful of hunters.

One evening, the lights had been lit and I was waiting patiently to see what they were going to add to my collection; it was still fairly early, so most of the predators, apart from myself and a few bats, had not put in an appearance. The bats flew up and down the verandah as fast as whip-lashes, taking the moths and other succulent dainties from within inches of the lamp, the wind from their wings making the flames shudder and leap. Gradually, the pale dragon-green afterglow of the sunset faded, the crickets started their prolonged musical trills, the gloom of the olive trees was lit by the cold lights of the fireflies, and the great house, creaking and groaning with sunburn, settled down for the night.

The wall behind the lamp was already covered by a host of various insects which, after a preliminary unsuccessful suicide attempt, were clinging there to recover themselves before trying again. At the base of the wall, from a minute crack in the plaster, emerged one of the smallest and fattest geckoes I had ever seen. He must have been newly hatched, for he measured only about an inch and a half in length, but obviously the short time he had been in the world had not prevented him from eating prodigiously, for his body and tail were so fat as to make him appear almost circular. His mouth was set in a wide, shy smile, and his large dark eyes were wide and wondering, like the eyes of a child that sees a table set for a banquet. Before I could stop him, he had waddled slowly up the wall and started his supper with a lacewing fly; these creatures, with their transparent wings like green lace and their large green-gold eyes, were favorites of mine and so I was annoyed with him.

Gulping down the last bit of gauzy wing, the baby gecko paused, clinging to the wall, and mused for a bit, occasionally blinking his eyes. I could not think why he had chosen the lacewing, which was a fairly bulky thing to handle, when he was surrounded on all sides by a variety of small insects which would have been much easier for him to catch and eat. But it soon became apparent that he was a glutton whose eyes were bigger than his stomach. Having hatched from an egg – and, therefore, lacking a mother's guidance – he was under the strong but erroneous impression that all insects were edible and that the bigger they were the quicker they would assuage his hunger. He did not even seem to be aware of the fact that for a creature of his size some insects could be dangerous. Like an early missionary, he was so concerned with himself that it never occurred to him that somebody might look upon him simply as a meal.

Ignoring a convention of small and eminently edible moths sitting near him, he stalked a great, fat, hairy oak eggar whose body was almost bigger than his own; he misjudged his run-in, however, and merely caught her by the tip of one wing. She flew off, and such was the power of her brown wings that she almost tore the gecko's grip from the wall and carried him with her. Nothing daunted, after a brief rest the gecko launched an assault on a longicorn beetle his own size. Quite apart from anything else, he would never have been able to swallow such a hard, prickly monster, but this apparently did not occur to him. However, he could not get a grip on the beetle's hard and polished body, and all he succeeded in doing was knocking it to the floor.

He was just having another brief rest and surveying the battlefield when, with a great crisp rustle of wings, an enormous mantis flew onto the verandah and alighted on the wall some six inches away from the gecko. She folded her wings with a noise like the crumpling of tissue paper, and with her viciously pronged arms raised in mock prayer she stared about her with her lunatic eyes, twisting her head from side to side as she surveyed the array of insects assembled for her benefit. The gecko, it was fairly obvious, had never seen a mantis before and did not realize how lethal they could be; as far as he was concerned, it was an enormous green dinner of the sort that he had dreamed about but never hoped to obtain, Without more ado, and ignoring the fact that the mantis was some five times his size, he commenced to stalk her. The mantis, meanwhile, had singled out a silver-Y moth and was moving towards it on her attenuated, elderly-spinster legs, pausing occasionally to sway to and fro, the personification of evil. Hard in her wake came the gecko, head down, grimly determined, pausing whenever the mantis did and lashing his ridiculous little fat tail to and fro like an excited puppy. The mantis reached the oblivious moth, paused, swaying, and then lashed out with her foreclaws and seized it. The moth, which was a large one, started fluttering frantically, and it required all the strength of the mantis's cruelly barbed forelegs to hold it. As she was struggling with it, looking like a rather inept juggler, the gecko, who had lashed himself into a fury with his fat tail, launched his attack. He darted forward and laid hold of the mantis's wing case like a bulldog. The mantis was busy trying to juggle the moth around in her claws, so this sudden attack from the rear got her off balance and she fell to the ground, carrying with her the moth and the gecko. When she landed, she still had the gecko hanging grimly to her wing case, so she relinquished the moth, which was by now almost dead, to leave her sabre-sharp front claws free to do battle with the gecko.

I had just decided that this was the point where I should step in and obtain the moth for my collection and a mantis and a gecko to add to my menagerie when another protagonist entered the arena. From the shadows of the grapevine a scutiger slid into view, a moving carpet of legs, skimming purposefully towards the still twitching moth. It reached it, poured itself over the moth, and sank its jaws into the moth's soft thorax. It was now a fascinating scene; the mantis was bent almost double, slashing downwards with her needle-sharp claws at the gecko, who, with eyes protruding with excitement, was hanging on grimly though he was being whipped to and fro by his large antagonist. The scutiger, deciding it could not move the moth, lay draped over it like a pelmet, sucking its vital juices out.

It was at that point that Theresa Olive Agnes Deirdre, known as Deirdre for short, made her appearance. Deirdre was one of a pair of enormous common toads I had found, tamed with comparative ease, and established in the tiny walled garden below the verandah. Here they lived a blameless life among the geraniums and tangerine trees, venturing up onto the verandah when the lights were lit to take their share of the insect life thus provided. So taken up was I by the strange foursome in front of me that I had forgotten all about Deirdre, so when she appeared on the scene I was unprepared, lying as I was on my stomach with my nose some six inches from the battlefield. Unbeknownst to me, Deirdre had been watching the skirmishing from beneath a chair. She now hopped forward fatly, paused for a brief second, gulping; then, before I could do anything, she leapt forward in the purposeful way that toads have, opened her huge mouth, and with the aid of her tongue flipped both the scutiger and the moth into her capacious maw. She paused again, gulping so that her protuberant eyes disappeared briefly, and then turned smartly to the left and flipped both mantis and gecko into her mouth. Only for a moment did the gecko's tail protrude, wriggling like a worm between Deirdre's thick lips, before she stuffed it firmly into her mouth, toad-fashion, with her thumbs.

I had read about food chains and the survival of the fittest, but this I felt was carrying things too far. Apart from anything else, I was annoyed with Deirdre for spoiling what was proving to be an absorbing drama. So that she would not interfere with anything else, I carried her back to the walled garden to the home she shared with her husband, Terence Oliver Albert Dick, under a stone trough full of marigolds. I reckoned she had eaten quite enough for one evening anyway.

So it was to a house baked crisp as a biscuit, hot as a baker's oven, and teeming with animal life that Adrian Fortesque Smythe made his appearance. Adrian, a school friend of Leslie's, had spent one holiday with us in England and as a result had fallen deeply and irrevocably in love with Margo, much to her annoyance. We were all spread out on the verandah reading our fortnightly mail when the news of Adrian's imminent arrival was broken to us by Mother.

"Oh, how nice," she said. "That will be nice."

We all stopped reading and looked at her suspiciously.

"*What* will be nice?" asked Larry.

"I've had a letter from Mrs. Fortesque Smythe," said Mother.

"I don't see anything nice about *that*," said Larry.

"What does the old hag want?" Leslie inquired.

"Leslie, dear, you mustn't call her an old hag. She was very kind to you, remember."

Leslie grunted derisively.

"What's she want anyway?" he asked.

"Well, she says Adrian's doing a tour of the Continent and could he come to Corfu and stay with us for a bit."

"Oh, good," said Leslie, "it'll be nice to have Adrian to stay."

"Yes, he's a nice boy," said Larry magnanimously.

"Yes, isn't he?" said Mother enthusiastically. "Such nice manners."

"Well, *I'm* not pleased he's coming," said Margo. "He's one of the most boring people I know. He makes me yawn just to look at him. Can't you say we're full up, Mother?"

"But I thought you liked Adrian," said Mother, surprised. "He certainly liked you, if I remember."

"That's just the point," said Margo. "I don't want him drooling all over the place like a sex-starved spaniel."

Mother straightened her spectacles and looked at Margo.

"Margo, dear, I don't think you ought to talk about Adrian like that," she said. "I don't know where you get these expressions. I'm sure you're exaggerating. I never saw him look like a . . . like a . . . well . . . like what you said. He seemed perfectly well behaved to me."

"Of course he was," said Leslie belligerently. "It's just Margo; she thinks every man is after her."

"I don't," said Margo indignantly. "I just don't like him. He's squishy. Every time you looked around, there he was, dribbling."

"Adrian never dribbled in his life."

"He did. Nothing but dribble, dribble, drool."

"I can't say *I* ever saw him dribbling," said Mother vaguely, "but I can't say he can't stay because he dribbles, Margo. Do be reasonable."

"He's Les's friend. Let him dribble over Les."

"He doesn't dribble. He's never dribbled."

"Well," said Mother, with the air of one solving a problem. "There'll be plenty for him to do, so I dare say he won't have time to dribble."

A fortnight later a starving, exhausted Adrian arrived, having cycled with practically no money all the way from Calais on a bicycle, which had given up the unequal struggle and fallen to bits in Brindisi. For the first few days we saw little of him, since Mother insisted he go to bed early, get up late, and have another helping of everything. When he did put in an appearance, I watched him narrowly for signs of dribbling, for of all the curious friends we had had staying with us, we had never had one that dribbled before and I was anxious to witness this phenomenon. But apart from a tendency to go scarlet every time Margo entered the room and to sit looking at her with his mouth slightly open (when, honesty compelled me to admit, he did look rather like a spaniel), he betrayed no other signs of eccentricity. He had extravagantly curly hair, large, very gentle hazel eyes, and his hormones had just allowed him to achieve a hairline moustache of which he was extremely proud. He had bought, as a gift for Margo, a record of a song which he obviously considered to be the equivalent of Shakespearean sonnets set to music. It was called "At Smokey Joe's," and we all grew to hate it intensely, for Adrian's day was not complete unless he had played this cacophonous ditty at least twenty times.

"Dear God," Larry groaned at breakfast one morning as he heard the hiss of the record, "not *again*, not at this hour."

"At Smokey Joe's in Havana," the gramophone proclaimed loudly in a nasal tenor voice, "I lingered quenching my thirst. . . ."

"I can't bear it. Why can't he play something else?" Margo wailed.

"Now, now, dear. He likes it," said Mother placatingly.

"Yes, and he bought it for *you*," said Leslie. "It's your bloody present. Why don't you tell him to stop?"

"No, you can't do that, dear," said Mother. "After all, he is a guest."

"What's that got to do with it?" snapped Larry. "Just because he's tone-deaf, why should we all have to suffer? It's Margo's record. It's her responsibility."

"But it seems so impolite," said Mother worriedly. "After all, he brought it as a present; he thinks we like it."

"I know he does. I find it hard to credit such depths of ignorance," said Larry. "D'you know he took off Beethoven's Fifth yesterday *halfway* through to put on that emasculated yowling! I tell you he's about as cultured as Attila the Hun."

"Sshh, he'll hear you, Larry dear," said Mother.

"What, with that row going on? He'd need an ear trumpet."

Adrian, oblivious to the family's restiveness, now joined the recorded voice to make a duet. As he had a nasal tenor voice remarkably similar to the vocalist's, the result was pretty horrible.

"I saw a damsel there . . . That was really where . . . I saw her first . . . Oh, Mama Inez . . . Oh, Mama Inez . . . Oh, Mama Inez . . . Mama Inez. . . ." warbled Adrian and the gramophone more or less in unison.

"God in heaven!" Larry exploded. "That's really too much! Margo, you've got to speak to him."

"Well, do it politely, dear, said Mother. "We don't want to hurt his feelings."

"I feel just like hurting his feelings," said Larry.

"I know," said Margo, "I'll tell him Mother's got a headache."

"That will only give us a temporary respite," said Larry.

"You tell him Mother's got a headache and I'll hide the needle," said Leslie triumphantly. "How about that?"

"Oh, that's a brain wave," Mother exclaimed, delighted that the problem had been solved without hurting Adrian's feelings.

Adrian was somewhat mystified by the disappearance of the needles and the fact that everyone assured him they could not be obtained in Corfu but could only be procured from England. However, he had a retentive memory, if no ability to carry a tune, so he hummed "At Smokey Joe's" all day long, sounding like a hive of very distraught tenor bees.

As the days passed, his adoration of Margo showed no signs of abating; if anything, it grew worse, and Margo's irritation grew with it. I began to feel very sorry for Adrian, for it seemed that nothing he could do was right. Because Margo said she thought his moustache made him look like an inferior gentlemen's hairdresser, he shaved it off, only to have Margo proclaim that moustaches were a sign of virility. Furthermore, she was heard to say in no uncertain terms that she much preferred the local peasant boys to any English import.

"They're so handsome and so sweet," she said, to Adrian's obvious chagrin. "They all sing so well. They have such nice manners. They play the guitar. Give me one of them to an Englishman any day. They have a sort of ordure about them."

"Don't you mean aura?" asked Larry.

"Anyway," Margo continued, ignoring this, "they're what I call *men*, not namby-pamby dribbling washouts."

"Margo, dear," said Mother, glancing nervously at the wounded Adrian, "I don't think that's very kind."

"I'm not trying to be kind," said Margo, "and most of cruelty is kindness if it's done in the right way."

Leaving us with this baffling piece of philosophy, she went off to see her latest conquest, a richly tanned fisherman with a luxuriant moustache. Adrian was so obviously mortified that the family felt it must try and alleviate his mood of despair.

"Don't take any notice of Margo, Adrian dear," said Mother soothingly. "She doesn't mean what she says. She's very headstrong, you know. Have another peach."

"Pigheaded," said Leslie. "And *I* ought to know."

"I don't see *how* I can be more like the peasant boys," said Adrian, puzzled. "I suppose I could take up the guitar."

"No, no, don't do that," said Larry hastily. "That's quite unnecessary. Why not try something simple? Try chewing garlic."

"Garlic?" said Adrian, surprised. "Does Margo like garlic?"

"Sure to," said Larry. "You heard what she said about

those peasant lads' auras. Well, what's the first bit of their aura that hits you when you go near them? Garlic!"

Adrian was much struck by the logic of this and chewed a vast quantity of garlic, only to be told by Margo, with a handkerchief over her nose, that he smelt like the local bus on market day.

Adrian seemed to me to be a very nice person; he was gentle and kind and always willing to do anything that anyone asked of him, so I found Margo's attitude very puzzling. I felt it my duty to try and do something for Adrian, but short of locking Margo in his bedroom – a thought which I dismissed as being impractical and an action liable to be frowned on by Mother – I could think of nothing very sensible. I decided to discuss the matter with Mr. Kralefsky in case he could suggest anything. I told him about Adrian's unsuccessful pursuit of Margo when we were having our coffee break, a welcome respite for us both from the insoluble mysteries of the square of the hypotenuse.

"Aha!" he said. "The paths of love never run smooth. One is tempted to wonder, indeed, if life would not be a trifle dull if the road to one's goal were always smooth."

I was not particularly interested in my tutor's philosophical flights but I waited politely. Mr. Kralefsky picked up a biscuit delicately in his beautifully manicured hands, held it briefly over his coffee cup and then christened it gently in the brown liquid before popping it into his mouth. He chewed methodically, his eyes closed.

"It seems to me," he said at last, "that this young Lochinvar is trying too hard."

I said that Adrian was English, but in any case, how could one try too hard? If one didn't try hard, one didn't achieve success.

"Ah," said Mr. Kralefsky archly, "but in matters of the

heart things are different. A little bit of indifference sometimes works wonders."

He put his fingertips together and gazed raptly at the ceiling, and I could tell that we were about to embark on one of his flights of fancy with his favorite mythological character, "a lady."

"I remember once I became greatly enamored of a certain lady," said Kralefsky. "I tell you this in confidence, of course."

I nodded and helped myself to another biscuit. Kralefsky's stories were apt to be a bit lengthy.

"She was a lady of such beauty and accomplishments that every eligible man flocked round her like . . . like . . . bees round a honey pot," said Mr. Kralefsky, pleased with this image. "From the moment I saw her I fell deeply, irrevocably, inconsolably in love with her, and I felt that she in some measure returned my regard."

He took a sip of coffee to moisten his throat, then he trellised his fingers together and leaned across the desk, his nostrils flaring, his great, soulful eyes intense.

"I pursued her relentlessly as a . . . as a . . . hound on the scent, but she was cold and indifferent to my advances. She even mocked the love that I offered her."

He paused, his eyes full of tears, and blew his nose vigorously.

"I cannot describe to you the torture I went through, the burning agony of jealousy, the sleepless nights of pain. I lost twenty-four kilos in weight; my friends began to worry about me, and, of course, they all tried to persuade me that the lady in question was not worthy of my suffering. All except one friend . . . a . . . an experienced man of the world, who had, I believe, had several affairs of the heart himself, one as far away as Baluchistan, and he told me that I was trying too hard, that as long as I was casting my heart at the lady's feet she was, like all females, content with her conquest. But if I showed a little indifference, aha! my friend assured me, it would be a very different tale."

Kralefsky beamed at me and nodded his head knowingly. He poured himself out more coffee.

And had he shown indifference, I asked.

"Indeed I did," said Kralefsky. "I didn't lose a minute. I embarked on a boat for China."

I thought this was splendid; no woman, I felt, could claim to have you enslaved if you suddenly leapt on a boat for China. It was sufficiently remote to give the vainest woman pause for thought. And what happened, I inquired eagerly, when Mr. Kralefsky returned from his travels.

"I found she had married," said Mr. Kralefsky, rather shamefacedly, for he obviously realized that this was somewhat of an anticlimax.

"Some women are capricious and impatient, you know, but I managed to have a few moments of private conversation with her and she explained it all."

I waited expectantly.

"She said," Mr. Kralefsky continued, "that she had thought I had gone for good to become a lama, so she married. Yes, the little dear would have waited for me had she known, but, torn with grief, she married the first man who came along. If I had not misjudged the length of the voyage, she would have been mine today."

He blew his nose violently, a stricken look on his face. I digested this story, which seemed to me highly suspect, but it did not seem to give me any clues as to how to help Adrian. Should I perhaps lend him my boat, the *Bootle Bumtrinket*, and suggest that he row over to Albania?

Apart from the risk of losing my precious boat, I did not think that Adrian was strong enough to row that far. No, I agreed with Kralefsky that Adrian was being too eager, but knowing how capricious my sister was, I felt she would greet Adrian's disappearance from the island with delight rather than with despair. I felt that Adrian's real difficulty lay in the fact that he could never get Margo alone. I decided that I would have to take Adrian in hand if he were going to achieve anything like success. The first thing was for him to stop following Margo around like a lamb following a sheep and to feign indifference, so I inveigled him into accompanying me when I went out to explore the surrounding countryside. This was easy enough to do. Margo, in self-defense, had taken to rising at dawn and disappearing from the villa before Adrian put in an appearance, so he was left pretty much to himself. Larry and Leslie were occupied with their own affairs. Mother tried to interest Adrian in cooking, but after he had left the icebox open and melted half our perishable foodstuffs, set fire to a frying pan full of fat, turned a perfectly good joint of lamb into something closely resembling biltong, and dropped half a dozen eggs onto the kitchen floor, Mother was only too glad to back up my suggestion that he accompany me.

To my surprise, I found Adrian an admirable companion, considering that he had been brought up in a city. He never complained, he would patiently obey my terse instructions – "Hold that!" or "Don't move!" or "Don't move, it'll bite you!" – to the letter, and he seemed genuinely interested in the creatures we pursued.

As Mr. Kralefsky had predicted, Margo became rather intrigued by Adrian's sudden absence. Although she did not care for his attentions, she felt perversely piqued when she was not receiving them. She wanted to know what Adrian and I did all day long, and I replied rather austerely that Adrian was helping me in my zoological investigations. I said that moreover he was shaping up very well, and that if this went on I would have no hesitation in proclaiming him a very competent naturalist by the end of the summer.

"I don't know how you can go around with anyone so wet," she said. "I find him an incredible bore."

I said that was probably just as well, as Adrian had confessed to me that he was finding Margo a bit boring too.

"What?" said Margo, outraged. "How dare he say that, how dare he!"

Well, I pointed out philosophically, she had only herself to blame. After all, who would not find someone boring if they carried on like she did, never going swimming with him, never going walking with him, always being rude. "I'm not rude," said Margo angrily. "I just speak the truth. And if he wants a walk, I'll give him one. Boring indeed!"

I was so pleased with the success of my scheme that I overlooked the fact that Margo, like the rest of my family, could be a powerful antagonist when aroused. That evening she was so unexpectedly polite and charming to Adrian that everyone, with the exception of Adrian, was amazed and alarmed. Skillfully, Margo steered the conversation round to walks and then said that as Adrian's time on the island was growing short, it was essential that he see more of it. And what better method than walking? Yes, stammered Adrian, that was really the best way of seeing a country.

"I intend to go for a walk the day after tomorrow," said Margo airily. "A lovely walk. It's a pity you're so busy with Gerry; otherwise you could have come with me."

"Oh, don't let that worry you. Gerry can fend for

himself," said Adrian, with what I privately considered to be callous and impolite indifference. "I'd love to come!"

"Oh, good," fluted Margo. "I'm sure you'll enjoy it; it's one of the nicest strolls around here."

"Where?" inquired Leslie.

"Liapades," said Margo airily. "I haven't been there for ages."

"Liapades?" echoed Leslie. "A stroll? It's right the other side of the island. It'll take you hours."

"Well, I thought we'd take a picnic and make a day of it," said Margo, adding archly, "that is, if Adrian doesn't mind."

It was quite obvious that Adrian would not mind if Margo had suggested swimming underwater to Italy and back in full armor. I said I thought I would accompany them as it was an interesting walk from a zoological point of view. Margo shot me a baleful look.

"Well, if you come you must behave yourself," she said enigmatically.

Adrian was, needless to say, full of the walk and Margo's kindness in asking him. I was not so sure. I said that Liapades was a long way and that it was very hot, but Adrian said he did not mind a bit. Privately, I wondered, since he was rather frail, whether he would last the pace, but I could not say this to him without insulting him. So at five o'clock on the appointed day we assembled on the verandah. Adrian was wearing an enormous pair of hobnailed boots he had acquired from somewhere, long trousers and a thick flannel shirt. To my astonishment, when I ventured to suggest that this ensemble was not suitable for a walk across the island in a temperature of over a hundred in the shade, Margo disagreed. Adrian was wearing perfect walking kit, chosen by herself, she said. The fact that she was clad in a diaphanous bathing suit and sandals and I was in shorts and an open-necked shirt did not deter her. She was armed with a massive pack on her back, which I imagined contained our food and drink, and a stout stick. I was carrying my collecting bag and butterfly net.

Thus equipped, we set out, Margo setting an unreasonably fast pace, I thought. Within a remarkably short space of time Adrian was sweating profusely, and his face turned pink. Margo, in spite of my protests, stuck to open country and shunned the shade-giving olive groves. In the end I kept pace with them, but walking in the shade of the trees a few hundred yards away. Adrian, afraid of being accused of being soft, followed doggedly and moistly at Margo's heels. After four hours, Adrian was limping badly and dragging his feet; his grey shirt was black with sweat and his face was an alarming shade of magenta.

"Would you like a rest?" Margo inquired at this point.

"Just a drink, perhaps," said Adrian in a parched voice like a corncrake.

I said I thought this was a splendid idea, so Margo stopped and sat down on a red-hot rock in the open sun-soaked ground on which you could have roasted a team of oxen. She fumbled surreptitiously in her pack and produced three small bottles of Gazoza, a fizzy and extremely sweet local lemonade.

"Here," she said, handing us a bottle each, "this'll buck you up."

In addition to being fizzy and oversweet, the Gazoza was very warm, so if anything it increased, rather than assuaged, our thirst. By the time it was nearing midday, we were in sight of the opposite coast of the island. A spark of hope crept into the lackluster eyes of Adrian. Once we reached the sea we could rest and swim, Margo explained. We reached the wild coastline and made our way down through the jumble of gigantic red and brown rocks strewn along the seashore like an uprooted giants' cemetery. Adrian threw himself down in the shade of an enormous block of rock topped with a wig of myrtle and a baby umbrella pine and tore off his shirt and boots. His feet, we discovered, were almost the same startling red as his face, and badly blistered. Margo suggested that he soak them in a rock pool to harden them, and this he did while Margo and I swam. Then, much refreshed, we squatted in the shade of the rocks, and I said I thought some food and drink would be welcome.

"There is none," said Margo.

There was a stunned silence for a moment.

"What d'you mean, there is none?" asked Adrian. "What's in that pack?"

"Oh, those are just my bathing things," said Margo. "I decided I wouldn't bring any food, because it was so heavy to carry in this heat, and anyway, we'll be back for supper if we start soon."

"And what about something to drink?" inquired Adrian hoarsely. "Haven't you got any more Gazoza?"

"No, of course not," said Margo irritably. "I brought *three*. That's one each, isn't it? And they're terribly heavy to carry. I don't know what you're fussing about anyway; you eat far too much. A little rest'll do you good. It'll give you a chance to unbloat."

Adrian came as close to losing his temper as I had ever seen him.

"I don't *want* to unbloat, whatever that means," he said icily, "and if I did, I wouldn't walk half across the island to do it."

"That's just the trouble with you. You're namby-pamby," snorted Margo. "Take you for a little walk and you're screaming for food and wine. You just want to live in the hub of luxury all the time."

"I don't think a drink on a day like this is a luxury," said Adrian. "It's a necessity."

Finding this argument profitless, I took the three empty Gazoza bottles half a mile down the coast to where I knew there was a tiny spring. When I reached it, I found a man squatting by it, having his midday meal. He had a brown, seamed, wind-patterned face and a sweeping black moustache. He was wearing the thick, sheep's-wool socks that the peasants wear when working in the fields, and beside him lay his short, wide-bladed hoe.

"*Kalimera*," he greeted me without surprise, and waved his hand in a courteous gesture towards the spring, as if he owned it.

I greeted him and then lay face downwards on the small carpet of green moss that the moisture had created. I lowered my face to where the bright spring throbbed like a heart under some maidenhair ferns. I drank long and deeply and I could never remember water having tasted so good. I soaked my head and neck with it and sat up with a satisfied sigh.

"Good water," said the man. "Sweet, huh? Like a fruit."

I said the water was delicious and started to wash the Gazoza bottles and fill them.

"There's a spring up there," said the man, pointing up the precipitous mountainside, "but the water is different, bitter as a widow's tongue. But this is sweet, kind water. You are a foreigner?"

While I filled the bottles I answered his questions, but my mind was busy with something else. Nearby lay the remains of his food – half a loaf of maize bread, yellow as a primrose, some great fat white cloves of garlic and a handful of large, wrinkled olives as black as beetles. At the sight of them my mouth started to water, and I became acutely aware of the fact that I had been up since dawn with nothing to eat. Eventually the man noticed the glances I kept giving his food supply, and with the typical generosity of the peasants, he pulled out his knife.

"Bread?" he asked. "You want bread?"

I said that I would love some bread but that the problem was that there were three of me, as it were. My sister and her husband, I lied, were also starving somewhere among the rocks. He snapped his knife shut, gathered together the remains of his lunch and held it out to me.

"Take it for them," he said, grinning. "I've finished, and it's not right for the good name of Corfu that foreigners should starve."

I thanked him profusely, put the olives and garlic into my handkerchief, tucked the bread and the Gazoza bottles under my arm and set off.

"Go to the good," the man called after me. "Keep away from the trees. We'll be having a storm later."

Looking up at the blue and burnished sky, I thought the man was wrong but did not say so. When I got back, I found Adrian sitting glumly with his feet in a rock pool and Margo sunbathing on a rock and singing tunelessly to herself. They greeted the food and water with delight and fell on it, tearing at the golden bread and gulping the olives and garlic like famished wolves.

"There," said Margo brightly when we had finished, as if she had been responsible for providing the viands, "that was nice. Now I suppose we'd better be getting back."

Immediately a snag became apparent. Adrian's feet, cool and happy from the rock pool, had swollen. It took the united efforts of Margo and me to get his boots on again, and once we had succeeded in forcing his feet into the boots, he could only progress at a painfully slow pace, limping along like an elderly tortoise.

"I do wish you'd hurry up," shouted Margo irritably after we had progressed a mile or so and Adrian was lagging behind.

"I can't go any faster. My feet are killing me," Adrian said miserably.

In spite of our protests, that he would get sunburned, he had taken off his flannel shirt and exposed his milk-white skin to the elements. When we were a couple of miles from the villa, the peasant's prophecy about the storm became fact. These summer storms hatched in a nest of cumulus clouds in the Albanian mountains and were ferried rapidly across to Corfu by a warm, scouring wind like the blast from a baker's oven. The wind hit us now, stinging our skins and blinding us with dust and bits of leaf. The olives changed from green to silver, like the sudden gleam of a turning school of fish, and the wind roared its way through a million leaves with a noise like a giant breaker on the shore. The blue sky was suddenly, miraculously blotted out by bruise-colored clouds that were splintered by jagged spears of lavender lightning. The hot, fierce wind increased, and the olive groves shook and hissed as though shaken by some huge invisible predator. Then came the rain, plummeting out of the sky in great gouts, hitting us with the force of stones from a slingshot. The background to all this was the thunder, stalking imperiously across the sky, rumbling and snarling above the scudding clouds like a million stars colliding, crumbling and avalanching through space.

This was one of the best storms we had ever experienced, and Margo and I were thoroughly enjoying it, for after the heat and stillness we found the stinging rain and the noise exhilarating. Adrian did not share our view; he was one of those unfortunate people who are terrified of storms, so to him the whole thing was monstrous and frightening. We tried to take his mind off the storm by singing, but the thunder was so loud that he could not hear us. We struggled on grimly and at last, through the gloomy, rain-striped olive groves we saw the welcoming lights of the villa. As we reached it and Adrian staggered in through the front door, seeming more dead than alive, Mother appeared in the hall.

"Where *have* you children been? I was getting quite worried," she said. Then, catching sight of Adrian, she said, "Good heavens, Adrian dear, what *have* you been doing?"

She might well have asked, for those parts of Adrian's anatomy that were not scarlet with sunburn were interesting shades of blue and green; he could hardly walk, and his teeth were chattering so violently that he could not talk. Being scolded and commiserated with in turns, he was whisked away to bed by Mother. There he lay, with mild sunstroke, a severe cold and septic feet, for the next few days.

"Really, Margo, you do make me angry sometimes," said Mother. "You know he's not strong. You might have killed him."

"Serves him jolly well right," said Margo callously. "He shouldn't have said I was boring. It's an eye for an ear."

Adrian, however, unwittingly got his own back; when he recovered, he found a shop in the town that stocked gramophone needles.

Dogs, Dormice and Disorder

The unspeakable Turk should immediately be struck out of the question.

CARLYLE

THAT SUMMER was a particularly rich one; it seemed as if the sun had drawn up a special bounty from the island, for never had we had such an abundance of fruit and flowers, never had the sea been so warm and filled with fish, never had so many birds reared their young, or butterflies and other insects hatched and shimmered across the countryside. Watermelons, their flesh as crisp and cool as pink snow, were formidable botanical cannonballs, each one big enough and heavy enough to obliterate a city; peaches as orange or pink as a harvest moon loomed huge in the trees, their thick, velvety pelts swollen with sweet juice; the green and black figs burst with the pressure of their sap, and in the pink splits the gold-green rose beetles sat dazed by the rich, never ending largesse of food. Trees had been groaning with the weight of cherries, and the orchards looked as though some great dragon had been slain among the trees, bespattering the leaves with scarlet and wine-red drops of blood. The maize cobs were as long as your arm, and as you bit into the canary-yellow mosaic of seeds, the white, milky juice burst into your mouth. In the trees, swelling and fattening themselves for autumn, were the jade-green almonds and walnuts, and olives, smoothly shaped, bright and shining as birds' eggs strung among the leaves.

Naturally, with the island thus aburst with life, my collecting activities redoubled. As well as my regular weekly afternoon spent with Theodore, I now undertook much more daring and comprehensive expeditions than I had been able to before, for now I had acquired a donkey. This beast, Sally by name, had been a birthday present, and as a means of covering long distances and carrying a lot of equipment I found Sally an invaluable, if stubborn, companion. To offset her stubbornness she had one great virtue, and that was that she was, like all donkeys, endlessly patient. She would gaze happily into space while I watched some creature or other, or else she would simply fall into a donkey doze, that happy, trancelike state that donkeys can attain when, with half-closed eyes, they appear to be dreaming of some nirvana and become impervious to shouts, threats, or even whacks with sticks. The dogs, after a short period of patience, would start to yawn and sigh and scratch and show many small signs that they felt we had devoted enough time to a spider or whatever it was and move on. Sally, however, once she was in her doze, gave the impression that she would happily stay there for several days if the necessity arose.

One day a peasant friend of mine, a man who had obtained a number of my specimens for me and who was a careful observer, informed me that there were two huge birds hanging about in a rocky valley some five miles north of the villa. He thought that they must be nesting there. From his description they could only be eagles or vultures, and I was most anxious to try and get some young of either of these birds. My collection of birds of prey now numbered three species of owl, a sparrow hawk, a merlin and a kestrel, so I felt the addition of an eagle or vulture would round it off, so to speak. Needless to say I did not vouchsafe my ambition to the family, as already the meat bill for my animals was astronomical. Apart from this I could imagine Larry's reaction to the suggestion of a vulture's being inserted into the house. When acquiring new pets, I always found it wiser to face him with a *fait accompli*, for once the animals were introduced to the villa, I could generally count on getting Mother and Margo on my side.

I prepared for my expedition with great care, making up loads of food for myself and the dogs, a good supply of Gazoza as well as the normal complement of collecting tins and boxes, my butterfly net and a large bag to put my eagle or vulture in. I also took Leslie's binoculars, which were of a higher magnification than my own. He, luckily, was not around for me to ask, but I felt sure he would happily have lent them to me if he had been. So having checked my equipment for the last time to make sure nothing was missing, I proceeded to festoon Sally with the various items. She was in a singularly sullen and recalcitrant mood, even by donkey standards, and annoyed me by deliberately treading on my foot and then giving me a sharp nip on the buttock when I bent down to pick up my fallen butterfly net. She took grave offense at the clout I gave her for this misbehavior, so we started this expedition barely on speaking terms. Coldly, I fixed her straw hat carefully over her furry, lily-shaped ears, whistled to the dogs and set off.

Although it was still early, the sun was hot and the sky clear burning blue, like the blue you get by scattering salt on a fire, blurred at the edges with heat haze. To begin with we made our way along the road, thick with white dust as clinging as pollen, and we passed many of my peasant friends on their donkeys, going to market or down to their fields to work. This inevitably held up the progress of the expedition, for good manners required that you pass the time of day with each one, that you gossip for the right length of time and perhaps accept a crust of bread, some dry watermelon seeds or a bunch of grapes as a sign of your love and affection. So when it was time to turn off the hot, dusty road and start climbing through the cool olive groves, I was loaded down with a variety of edible commodities, the largest of which was a watermelon, a generous present pressed upon me by Mama Agathi, a friend of mine that I had not seen for a week, an unconscionable length of time, during which she presumed I had been without food.

The olive groves were dark with shadows and as cool as a well after the glare of the road. The dogs went ahead as usual, foraging around the great pitted olive boles and occasionally chasing skimming swallows, barking vociferously at their audacity. Failing, as always, to catch one, they would then attempt to vent their wrath on some innocent sheep or vacant-faced chicken and would have to be sternly reprimanded. Sally, her previous sulkiness forgotten, stepped out at a good pace, one ear pricked forward and the other one backward so that she could listen to my singing and comments on the passing scene.

Presently we left the shade of the olives and climbed upwards through the heat-shimmered hills, making our way through thickets of myrtle bushes, small copses of holm oak and great wigs of broom. Here Sally's hoofs crushed the herbs underfoot and the warm air became redolent with the scent of sage and thyme. By midday, the dogs panting, Sally and I sweating profusely, we were high up among the gold and rust-red rocks of the central range. Far below us lay the sea, blue as flax. By half-past two, pausing to rest in the shade of a massive outcrop of stone, I was feeling thoroughly frustated. We had followed the instructions of my friend and had indeed found a nest, which to my excitement proved to be that of a griffon vulture. Moreover, the nest, perched on a rocky ledge, contained two fat and almost fully fledged youngsters at just the right age for adoption. The snag was that I could not reach the nest, either from above or below. After having spent a fruitless hour trying to kidnap the babies, I was forced, albeit reluctantly, to give up the idea of adding vultures to my collection of birds of prey. So we moved down the mountainside and stopped to rest and eat in the shade. While I ate my sandwiches and hard-boiled eggs, Sally had a light lunch of dry maize cobs and watermelon, and the dogs assuaged their thirst with a mixture of watermelon and grapes, gobbling the juicy fruit eagerly and occasionally choking and coughing as a melon seed got stuck. Because of their voraciousness and total lack of table manners, they had finished their lunch long before Sally or I had, and having reluctantly come to the conclusion that I did not intend to give them any more to eat, they left us and slouched down the mountainside to indulge in a little private hunting.

I lay on my tummy eating crisp, cool watermelon pink as coral and examined the hillside. Fifty feet or so below where I lay were the ruins of a small peasant house. Here and there on the hillside you could just discern the crescent-shaped, flattened areas which had once been the tiny fields of the farm. When the impoverished soil would no longer support maize or vegetables, the owner had moved away. The house had tumbled down and the fields had become overrun with weeds and myrtle. I was staring at the remains of the cottage, wondering who had lived there, when I saw something reddish moving through the thyme at the base of one of the walls.

Slowly I reached out for the field glasses and put them to my eyes. The tumbled mass of rocks at the base of the wall sprang into clear view, but for a moment I could not see what it was that had attracted my attention. Then, to my astonishment, from behind a clump of thyme appeared a lithe, tiny animal as red as an autumn leaf. It was a weasel, and to judge by its behavior, a young and rather innocent one. It was the first weasel I had seen on Corfu and I was enchanted by it. It peered about with a slightly bemused air and then stood up on its hind legs and sniffed the air vigorously. Apparently not smelling anything edible, it sat down and had an intensive and, from the look of it, very satisfying scratch. Then it suddenly broke off from its toilet and carefully stalked and attempted to capture a vivid canary-yellow brimstone butterfly. The insect, however, slipped out from under its jaws and flipped away, leaving the weasel snapping at thin air and looking slightly foolish. It sat up on its hind legs once more to see where its quarry had gone and, overbalancing, almost fell off the stone it was sitting on.

I watched it, entranced by its diminutive size, its rich coloring and its air of innocence. I wanted above all things to catch it and take it home with me to add to my menagerie, but I knew this would be difficult. While I was musing on the best method of achieving this result, a drama unfolded in the ruined cottage below. I saw a shadow like a Maltese cross slide over the low scrub, and then a sparrow hawk appeared, flying low and fast towards the weasel, who was sitting up on his stone sniffing the air and apparently unaware of his danger. I was just wondering whether to shout or clap my hands to warn him when he saw the hawk. With incredible speed he turned and leapt gracefully onto the ruined wall and disappeared into a crack between two stones that I would have thought would not have allowed the passage of a slowworm, let alone a mammal the size of the weasel. It was like a conjuring trick. One minute he had been sitting on his rock and the next he vanished into the wall like a drop of rainwater. The sparrow hawk checked with fanned tail and hovered briefly, obviously hoping the weasel would reappear. After a moment or so it got bored and slid off down the mountainside in search of less wary game. After a short time the weasel poked his little face out of the crack and, seeing the coast was clear, emerged cautiously. Then he made his way along the wall, and as though his squeeze into the wall to escape had given him the idea, he proceeded to investigate and disappear into every nook and cranny that existed between the stones. As I watched him, I was wondering how to make my way down the hill to throw my shirt over him before he became aware of my presence. In view of his expert vanishing trick when faced with the hawk, it was obviously not going to be easy.

At that moment he slid, sinuous as a snake, into a hole in the base of the wall, and from another hole a little higher up there emerged another animal, in a great state of alarm, which made its way along the top of the wall and disappeared into a crevice. I was greatly excited, for even with the brief glimpse I had got of it, I recognized it as a creature that I had tried for many months to track down and capture, a garden dormouse, probably one of the most attractive of the European rodents. It was about half the size of a fullgrown rat, with cinnamon-colored fur, brilliant white underparts, a long furry tail ending in a brush of black and white hair and a black band of fur beneath the ears, which ran across the eyes and made it look ridiculously as though it were wearing an old-fashioned burglar's mask.

I was now in something of a quandary, for there below me were two animals I dearly wanted to possess, one hotly pursuing the other and both of them exceedingly wary. If my attack was not well planned, I stood a good chance of losing both animals. I decided to tackle the weasel first, as he was the more mobile of the two, and I felt that the dormouse would not move from its new hole if it were left undisturbed. On reflection I decided that my butterfly net was a more suitable instrument of capture than my shirt, so armed with it I made my way down the hillside with the utmost caution, freezing immobile every time the weasel appeared out of the hole and looked around. Eventually I got to within a few feet of the wall without being detected. I tightened my grip on the long handle of my net and waited for the weasel to come out from the depths of the hole he was now investigating. When he did emerge, he did so with such suddenness that I was unprepared. He sat up on his hind legs and stared at me with interest untinged by alarm. I was just about to take a swipe at him with my net when the three dogs came crashing through the bushes, tongues lolling, tails wagging, as vociferously pleased to see me as if we had been separated for months. The weasel vanished. One minute he was sitting there, frozen with horror at this avalanche of dogs, and the next minute he was gone. Bitterly I cursed the dogs and banished them to the higher reaches of the mountainside, where they went to lie in the shade, hurt and puzzled at my bad temper. Then I set about the task of trying to capture the dormouse.

Over the years the mortar between the stones had grown frail and heavy winter rains had washed it away, so now, to all intents and purposes, the remains of the house were a series of dry stone walls. This maze of intercommunicating tunnels and caves formed the ideal hideout for any small animal. There was only one way to hunt for an animal in this sort of terrain and that was to take the wall to pieces, so rather laboriously this is what I started to do. After having dismantled a good section of it, I had unearthed nothing more exciting than a couple of indignant scorpions, a few wood lice and a young gecko, which fled, leaving his writhing tail behind him. It was hot and thirsty work, and after an hour or so I sat down in the shade of the as yet undismantled wall to have a rest.

I was just wondering how long it would take me to demolish the rest of the wall when, from a hole some three feet from me, the dormouse appeared. It scrambled up like a somewhat overweight mountain climber and then, having reached the top, it sat down on its fat bottom and commenced to wash its face with great thoroughness, totally ignoring my presence. I could hardly believe my luck.

Slowly and with great caution I maneuvered my butterfly net towards him, got it in position and then clapped it down suddenly. This would have worked perfectly if the top of the wall had been a flat and even surface, but it was not, so I could not press the rim of the net down hard enough. The result was that there was a gap, and to my intense annoyance and frustration, the dormouse, recovering from its momentary panic, squeezed out from under the net, galloped along the wall and disappeared into another crevice. However, this proved to be its undoing, for it had chosen a cul-de-sac, and before it had discovered its mistake, I had clamped the net over the entrance. The next thing was to get it out and into the bag without getting bitten. This was not easy, and before I had finished, it had sunk its exceedingly sharp teeth into the ball of my thumb, so that I, the handkerchief and the dormouse were all liberally bespattered with gore before I finally got it into the bag. Delighted with my success, I mounted Sally and rode home in triumph with my new acquisition.

On arrival at the villa, I carried the dormouse up to my room and housed it in a cage that had until recently been the home of a baby black rat, which had met an unfortunate end in the claws of my scops owl, Ulysses, who was of the opinion that all rodents had been created by a beneficent providence in order to fill his stomach. I therefore made quite sure that my precious dormouse could not escape and meet a similar fate. Once it was in the cage, I could examine it more closely. I discovered it was a female with a suspiciously large tummy, which led me to believe that she might be pregnant. After some consideration I called her Esmeralda (I had just been reading *The Hunchback of Notre Dame* and had fallen deeply in love with the heroine) and provided her with a cardboard box full of cotton waste and dried grass in which to have her family should she indeed be pregnant.

For the first few days she would leap at my hand like a bulldog when I went to clean her cage or feed her, but within a week she had tamed down and tolerated me, though still viewing me with a certain reserve. Every evening Ulysses, on his special perch above the window, would wake up and I would open the shutters so that he could fly off into the moonlit olive groves and hunt, only returning for his plate of mincemeat at about two in the morning. Once he was safely out of the way I could let

Esmeralda out of her cage for a couple of hours' exercise. She proved to be an enchanting creature with enormous grace in spite of her rotundity. She would take prodigious and breathtaking leaps from the cupboard onto the bed (where she bounced as if it were a trampoline) and from the bed to the bookcase or table, using her long tail with its bushy end as a balancing rod. She was vastly inquisitive and nightly would subject the room and its contents to a minute scrutiny, scowling through her black mask with whiskers quivering. I discovered that she had an overriding passion for large brown grasshoppers, and she would often come and sit on my bare chest, as I lay in bed, and scrunch these delicacies up. The result was that my bed always seemed to contain a prickly layer of wing cases, bits of leg and chunks of horny thorax, for she was a greedy and not particularly well-mannered feeder.

Then came the exciting evening when, after Ulysses had floated on silent wings into the olive groves and commenced to call *toink toink* after the manner of his kind, I opened the cage door and she would not come out but lurked inside the cardboard box and made angry chittering noises at me. When I tried to investigate her bedroom, she fastened onto my forefinger like a tiger and I had great difficulty in getting her to let go. Eventually I managed to get her off, and holding her firmly by the scruff of the neck, I investigated the box and found there, to my infinite delight, eight babies, each the size of a hazelnut and as pink as a cyclamen bud.

Delighted with Esmeralda's happy event, I showered her with grasshoppers, melon seeds, grapes and other delicacies of which I knew she was particularly fond, and followed the progress of the babies with breathless interest. Gradually their eyes opened and their fur grew. Within a short time the more powerful and adventurous of them would climb laboriously out of their cardboard nursery and wobble about on the floor of the cage when Esmeralda was not looking. This filled her with alarm, and she would pick this or that errant baby up in her mouth and, uttering peevish growling noises, transfer it to the safety of the bedroom. This was all very well with one or two, but as soon as all eight babies reached the inquisitive stage it was impossible for her to control them, so she had to let them wander at will. They started to follow her out of the cage, and it was then that I discovered that dormice, like shrews, have a habit of caravaning. That is to say that Esmeralda would go first; hanging onto her tail would be baby number one, hanging onto his tail would be baby number two, and onto his or her tail, baby number three, and so on. It was an enchanting sight to see these nine diminutive creatures, each wearing a little black mask, wending their way around the room like an animated furry scarf, flying over the bed or shillying up the table leg. A scattering of grasshoppers on the bed or floor, and the babies, squeaking excitedly, would gather round to feed, looking ridiculously like a convention of bandits.

Eventually, when the babies were fully adult, I was forced to take them into the olive grove and let them go, for the task of providing sufficient food for nine rapacious dormice was proving too time-consuming. I released them at the edge of the olive grove, near a thicket of holm oak, and there they colonized successfully. In the evenings, when the sun was setting and the sky was getting as green as a leaf, striped with sunset clouds, I used to go down to see the little masked dormice flitting through the branches with a ballerinalike grace, chittering and squeaking to each other as they pursued moths or fireflies or other delicacies through the shadowy branches.

IT WAS AS a result of one of my many forays on donkeyback that we got, as it were, infested by dogs. We had been up in the hills, where I had been endeavoring to catch some agamas on the glittering gypsum cliffs. We returned towards evening, when the shadows lay everywhere, charcoal-black, and everything was bathed in the slanting, soft golden light of the sinking sun. We were all hot and tired, hungry and thirsty, for we had long since eaten and drunk everything that we had brought with us. The last vineyard we had passed had yielded only some bunches of very black winegrapes whose sharp vinegariness had made the dogs curl back their lips and screw up their eyes and had left me feeling hungrier and thirstier than ever; so I decided that as leader of the expedition it was up to me to provide sustenance for the rest of the crew. I reined in and thought about the problem.

We were equidistant from three sources of food. There was the shepherd, old Yani, who would, I knew, give us cheese and bread, but his wife would probably still be in the fields and Yani himself might not have returned from grazing his goat flock. There was Agathi, who lived alone in a tiny, tumbledown cottage, but she was so poor that I always felt guilty about accepting anything from her, and, in fact, always made a point of sharing my food with her when I was around that way. Finally, there was sweet and gentle Mama Kondos, a widow of some eighty summers, who lived with her three unmarried and, as far as I could see, unmarriageable daughters in an untidy but prosperous farm in a valley to the south. They were quite well off by peasant standards, owning, apart from five or six acres of olives and agricultural land, two donkeys, four sheep and a cow. They were what one might call the landed gentry of the area, so I decided that the honor of revictualing my expedition would fall to them.

The three inordinately fat, ill-favored but good-natured girls had just returned from working in the fields and were gathered round the small well, bright and shrill as parrots, washing their fat, hairy brown legs. Mama Kondas herself, like a diminutive clockwork toy, was trotting to and fro scattering maize for the squawking, tousled flock of chickens. There was nothing straight about Mama Kondas; her little body was bent like a sickle blade, her legs were bowed with years of carrying heavy loads on her head, her arms and hands were permanently bent from picking things up; even her upper and lower lips curved inwards over her toothless gums, and her snow-white, dandelion-seed eyebrows curved over her blue-rimmed black eyes, which in their turn were guarded on each side by a fence of curved wrinkles in a skin as delicate as a baby mushroom's.

The daughters, on seeing me, gave shrill cries of joy and gathered round me like benign shire horses, clasping me to their mammoth bosoms and kissing me, exuding affection, sweat, and garlic in equal quantities. Mama Kondas, a small, bent David among these great aromatic Goliaths, beat them aside, shouting shrilly, "Give him to me, give him to me! My golden one, my heart, my love! Give him to me." She clasped me to her and covered my face with bruising kisses, for her gums were as hard as a tortoise's mouth.

At length, after I had been thoroughly kissed and patted and pinched all over to make sure I was real, I was allowed to sit down and offer some explanation as to why I had deserted them for so long. Did I not realize that it was a whole week since I had visited them? How could my love be so cruel, so tardy, so ephemeral? Still, since I was here at last, would I like some food? I said yes, I would love some, and some for Sally as well. The dogs, more ill-mannered, had helped themselves; Widdle and Puke had torn sweet white grapes off the vine that trailed over part of the house and were gulping them down greedily, while Roger, who appeared to be more thirsty than hungry, had gone beneath the fig and almond trees and disemboweled a watermelon and was lying with his nose stuck into its cool pink interior, his eyes closed in ecstasy, sucking the sweet icy juice through his teeth. Immediately, Sally was given three cobs of ripe corn to chew on and a bucket of water to slake her thirst, while I was presented with a mammoth sweet potato, its skin black and deliciously charcoaly from the fire, its sweet flesh beautifully soggy, a bowl of almonds, some figs, two enormous peaches, a hunk of yellow bread, olive oil and garlic.

Once I had engulfed this provender and thus taken the edge off my hunger, I could concentrate on exchanging gossip. Pepi had fallen out of an olive tree and broken his arm, silly boy; and Yani – no, not *that* Yani, the Yani over on the other side of the hill – had quarreled with Taki over the price of a donkey and Taki had got so angry he had fired his shotgun at the side of Yani's house, only it had been a very dark night and Taki was drunk and it had turned out to be Spiro's house, so now none of them were speaking. For some time we discussed the foibles and dissected the characters of our fellow men with great relish, and then I noticed that Lulu was missing from the scene. Lulu was Mama Kondas's dog, a lean, long-legged bitch with huge soulful eyes and long floppy ears like a spaniel. Like all peasant dogs, she was gaunt and scabby, her ribs sticking out like the strings of a harp, but she was an endearing dog and I was fond of her. Normally she was one of the first to greet me, and now she was nowhere to be seen. Had anything happened to her, I asked.

"Puppies!" said Mama Kondas. "*Po, po, po, po*. Eleven! Would you believe it?"

They had tied Lulu up to an olive tree near the house when the birth had become imminent, and she had crawled into the depths of the tree's trunk to have her young. After she had greeted me with enthusiasm, she watched with interest as I crawled into the olive tree on hands and knees and extracted the puppies to look at. As always, I was amazed that such scrawny, half-starved mothers could produce such plump, powerful puppies, with squashed, belligerent faces and loud sea-gull voices. They were, as usual, a wide variety of colors – black and white, white and tan, silver and bluish-grey, all black and all white. Any litter of Corfu puppies displayed such a wide variety of color schemes that to settle a question of paternity was virtually impossible. I sat with the mewling patchwork of puppies in my lap and told Lulu how clever she was, and she wagged her tail furiously.

"Clever, huh?" said Mama Kondos sourly. "Eleven puppies isn't clever, it's wanton. We shall have to get rid of all but one."

I was well aware that Lulu could not possibly be allowed to keep her full complement of puppies; in fact, she was lucky that they were going to leave her one. However, on this occasion I felt I might be of use to Lulu. I said that I felt sure that my mother would not only be delighted at the thought of having a puppy but said that she would be overwhelmed with gratitude to the Kondos family and Lulu for providing her with one. I therefore, after much thought, chose the one I liked best, a slug-fat, screaming little male who was black, white and grey with bright, corn-colored eyebrows and feet. I asked them to save this one for me until he was old enough to leave Lulu; in the meantime, I would apprise Mother of the exciting fact that we had acquired another dog, which would bring our full complement up to five, a nice round figure, I considered.

To my astonishment, Mother was not a bit pleased with the suggested increase in our dog tribe.

"No, dear," she said firmly, "we are not having another dog. Four is quite enough. And what with all your owls and everything, it's costing a fortune in meat anyway. No, I'm afraid another dog is out of the question."

In vain I argued that the puppy would be killed if we did not intervene. Mother remained firm. There was only one thing to be done; I had noticed in the past that Mother, faced with a hypothetical question like Would you like a nestful of baby redstarts? would say no firmly and automatically. Faced, however, with the nestful of baby birds, she would inevitably waver and then say yes. Obviously there was only one thing to be done, and that was to show her the puppy. I was confident that she would never be able to resist his golden eyebrows and socks, so I sent a message down to the Kondoses asking if I could borrow the puppy to show Mother, and one of the fat daughters obligingly brought it up the next day. But when I unwrapped it from the cloth in which she had brought it, I found to my annoyance that Mama Kondos had sent the wrong puppy. I explained this to the daughter, who said that she could do nothing about it as she was on her way in to the village, but I had better go and see her mother. She added that I had better make haste, as Mama Kondos had mentioned that she was intending to destroy the puppies that morning. Speedily, I mounted Sally and galloped through the olive groves.

When I reached the farm, I found Mama Kondos sitting in the sun stringing garlic heads together into white, knobbly plaits, while around her the chickens scratched and purred contentedly. After she had embraced me, asked after the health of me and my family and given me a plate of green figs, I produced the puppy and explained my errand.

"The wrong one?" she exclaimed, peering at the yelling puppy and prodding it with her forefinger. "The wrong one? How stupid of me. *Po po po po*, I quite thought it was the one with white eyebrows you wanted."

Had she, I inquired anxiously, destroyed the rest of the puppies?

"Oh, yes," she said absently, still staring at the puppy, "yes, this morning, early."

Well, then, I said resignedly, since I could not have the one I had set my heart on, I had better have the one she had saved.

"No, I think I can get you the one you want," she said, getting to her feet and fetching a broad-bladed hoe.

How, I wondered, could she get me my puppy if she had destroyed them? Perhaps she was going to retrieve the corpse for me, and I had no desire for that. I was just going to say so when Mama Kondos, mumbling to herself, trotted off to a field near the house, where the stalks of the first crop of maize stood yellow and brittle in the hot sun-cracked ground. Here she cast about for a moment and then started to dig. With the second sweep of the hoe she dug up three screaming puppies, their legs paddling frantically, ears, eyes and pink mouths choked with earth.

I was paralyzed with horror. She checked the puppies she had dug up, found they did not include the one that I wanted, threw them to one side, and recommenced digging. It was only then that the full realization of what Mama Kondos had done swept over me. I felt a great scarlet bubble of hate burst in my chest, and tears of rage poured down my cheeks. From my not incomprehensive knowledge of Greek insults, I dragged up the worst in my vocabulary. Yelling these at Mama Kondos, I pushed her out of the way so hard that she sat down suddenly, bewildered, among the cornstalks. Still calling down the curses of every saint and deity I could think of, I seized the hoe and rapidly but carefully dug up the rest of the gasping puppies. Mama Kondos was too astonished by my sudden change from calm to rage to say anything; she just sat there, openmouthed. I stuffed the puppies unceremoniously into my shirt, collected Lulu and the pup that had been left to her, and rode off on Sally, shouting curses over my shoulder at Mama Kondas, who had now got to her feet and was running after me shouting, "But, golden one, what's the matter? Why are you crying? You can have all the puppies. What's the matter?"

I burst into the house, hot, tear-stained, covered with mud, my shirt bulging with puppies. Lulu trotted at my heels, delighted with this sudden and unexpected outing for herself and her offspring. Mother was, as usual, embedded in the kitchen, making various delicacies for Margo, who had been away touring the mainland of Greece to recover from yet another unfortunate affair of the heart. Mother listened to my incoherent and indignant account of the puppies' premature burial and was shocked.

"Really!" she exclaimed indignantly, "these peasants!

I can't understand how they can be so cruel. Burying them alive! I never heard of such a barbarous thing. You did quite right to save them, dear. Where are they?"

I ripped open my shirt as though committing hara-kiri and a cascade of wriggling puppies fell out onto the kitchen table, where they started to grope their way blindly about, squeaking.

"Gerry, dear, not on the table where I'm rolling pastry," said Mother. "Really, you children! Yes, well, even if it's clean mud, we don't want it in the pies. Get a basket."

I got a basket and we put the puppies into it. Mother peered at them.

"Poor little things," she said. "There do seem to be an awful lot of them. *How* many? Eleven! Well, I don't know what we'll do with them. We can't have eleven dogs with the ones we've got."

I said hastily that I had got it all worked out; as soon as the puppies were old enough, we would find homes for them. I said that Margo, who would be home by then, could help me; it would be an occupation for her and keep her mind off sex.

"Gerry, dear!" said Mother, aghast, "don't say things like that. Whoever told you that?"

I explained that Larry had said that she needed her mind taken off sex and I thought that the puppies' arrival would achieve this happy result.

"Well, you mustn't say things like that," said Mother. "Larry's got no business to say things like that. Margo's just . . . just . . . a bit . . . emotional, that's all. Sex has nothing to do with it; that's something *quite* different. Whatever would people think if they heard you? You mustn't go around saying things like that. Now go and put the puppies somewhere safe."

So I took the puppies to a convenient olive tree near

the verandah and tied Lulu to it and cleaned the puppies up with a damp cloth. Lulu, deciding that baskets were very effete places in which to bring up puppies, immediately excavated a burrow between the friendly roots of the tree and carefully transferred her puppies to it one by one. To his annoyance, I spent more time cleaning up my special puppy than the others and tried to think of a name for him. Finally, I decided to call him Lazarus, Laz for short. I placed him carefully with his brothers and sisters and went to change my mud- and urine-stained shirt.

I arrived at the lunch table in time to hear Mother telling Leslie and Larry about the puppies.

"It's extraordinary," said Leslie. "I don't think they mean to be cruel; they just don't think. Look at the way they shove wounded birds into their game bags. So what happened? Did Gerry drown the puppies?"

"No, he did *not*," said Mother indignantly. "He brought them here, of course."

"Dear God!" said Larry. "Not more dogs! We've got four already."

"They're only puppies," said Mother, "poor little things."

"How many are there?" asked Leslie.

"Eleven," said Mother reluctantly.

Larry put down his knife and fork and stared at her. "Eleven?" he repeated. "Eleven? Eleven puppies! You must be mad."

"I keep telling you, they're only puppies – tiny little things," said Mother, flustered. "And Lulu's very good with them."

"Who the hell's Lulu?" asked Larry.

"Their mother. She's a dear," said Mother.

"So that's twelve bloody dogs."

"Well, yes, I suppose it is," said Mother. "I hadn't really counted."

"That's the trouble round here," snapped Larry. "Nobody counts! And before you know where you are, you're knee-deep in animals. It's like the bloody creation all over again, only worse. One owl turns into a battalion before you know where you are; sex-mad pigeons defying Marie Stopes in every room of the house; the place is so full of birds it's like a bloody poulterer's shop, to say nothing of snakes and toads and enough small fry to keep Macbeth's witches in provender for years. And on top of all that you go and get twelve more dogs. It's a perfect example of the streak of lunacy that runs in this family."

"Nonsense, Larry, you do exaggerate," said Mother. "Such a lot of fuss over a few puppies."

"You call eleven puppies a *few*? The place will look like the Greek branch of Cruft's Dog Show, and they'll all probably turn out to be bitches and *all* come into season simultaneously. Life will deteriorate into one long canine sexual orgy."

"Oh, yes," said Mother, changing the subject, "and you're not to go around saying Margo's sex-mad. People will get the wrong idea."

"Well, she is," said Larry. "I see no reason to cover up the truth."

"You know perfectly well what I mean," said Mother firmly. "I won't have you saying things like that. Margo's just romantic. There's a lot of difference."

"Well, all I can say is," said Larry, "when all those bitches you've brought into the house come into season together, Margo's going to have a lot of competition."

"Now, Larry, that's quite enough," said Mother, "and anyway, I don't think we ought to discuss sex at lunch."

A few days after this, Margo returned from her travels, bronzed and well and with her heart apparently healed. She talked incessantly about her trip and gave us graphic, if somewhat biased, thumbnail pictures of the people she had met, inevitably ending with "and so I told them if they came to Corfu to come and see us."

"I do hope you didn't invite everyone you met, Margo dear," said Mother, slightly alarmed.

"Oh, of course not, Mother," said Margo impatiently, having just told us about a handsome Greek sailor and his eight brothers to whom she had issued this lavish invitation. "I only asked the *interesting* ones. I would have thought you'd be glad to have some *interesting* people around."

"I get enough interesting people that Larry invites, thank you," said Mother acidly, "without you starting."

"This trip has opened my eyes," said Margo dramatically. "I've realized that you're all simply stagnating here. You're becoming narrow-minded and . . . and . . . insulated."

"I don't see that objecting to unexpected guests is being narrow-minded, dear," said Mother. "After all, I'm the one that has to do the cooking."

"But they're not unexpected," said Margo haughtily. "I invited them."

"Well, yes," said Mother, obviously feeling that she was not making much headway with this argument, "I suppose if they write and let us know they're coming, we can manage."

"Of course they'll let us know," said Margo coldly. "They're my friends; they wouldn't be so ill-mannered as not to let us know."

As it happened, she was wrong.

I returned to the villa after a very pleasant afternoon spent drifting up the coast in my boat looking for seals and, bursting, sun-glowing and hungry, into the drawing room in search of tea and the mammoth chocolate cake

I knew Mother had made, came upon a sight so curious that I stopped in the doorway, my mouth open in amazement, while the dogs, clustered around my legs, started to bristle and growl with astonishment. Mother was seated on the floor, perched uncomfortably on a cushion, gingerly holding in one hand a piece of rope to which was attached a small, black and excessively high-spirited ram. Sitting around Mother, cross-legged on cushions, were a fierce-looking old man in a tarbush and three heavily veiled women. Also ranged on the floor were lemonade, tea, and plates of biscuits, sandwiches and the chocolate cake. As I entered the room, the old man had just leaned forward, drawn a huge, heavily ornate dagger from his sash, and cut himself a large hunk of cake, which he was stuffing into his mouth with every evidence of satisfaction. It looked rather like a scene out of the *Arabian Nights*. Mother cast me an anguished look.

"Thank goodness you've come, dear," she said, struggling with the ram, who had gamboled into her lap by mistake. "These people don't speak English."

I inquired who they were.

"I don't know," said Mother desperately. "They just appeared when I was making tea; they've been here for hours. I can't understand a word they say. They insisted on sitting on the floor. I think they're friends of Margo's; of course, they may be friends of Larry's but they don't look highbrow enough."

I tried tentatively talking in Greek to the old man, and he leapt to his feet, delighted that someone understood him. He had a swooping, eagle nose, an immense white moustache like a frosty sheaf of corn, and black eyes that seemed to snap and crackle with his mood. He was wearing a white tunic with a red sash in which his dagger reposed, enormous baggy pants, long white cotton socks, and red, upturned *charukias* with immense pom-poms on the toes. So I was the adorable signorina's brother, was I, he roared excitedly, bits of chocolate cake trembling on his moustache as he talked. What an honor to meet me. He clasped me to him and kissed me so fervently that the dogs, fearing for my life, all started barking. The ram, faced with four vociferous dogs, panicked; it ran round and round Mother, twisting the rope around her, and then, at a particularly snarling bark from Roger, it uttered a frantic bleat and fled toward the French windows and safety, pulling Mother onto her back in a welter of spilt lemonade and chocolate cake. Things became confused.

Roger, under the impression that the old Turk was attacking Mother and me, launched an assault on the Turk's *charukias* and got a firm grip on one of the pom-poms. The old boy aimed a kick at Roger with his free foot and promptly fell down. The three women were sitting absolutely still, cross-legged on their cushions, screaming loudly behind their yashmaks. Mother's dog, Dodo, who had long ago decided that anything in the nature of a roughhouse was acutely distressing to a Dandie Dinmont of her lineage, merely sat soulfully in a corner and howled. The old Turk, who was surprisingly lithe for his age, had drawn his dagger and was making wild but ineffectual swipes at Roger, who was darting from pom-pom to pom-pom and growling savagely, evading the blade with ease. Widdle and Puke were trying to round up the ram, and Mother, desperately unraveling herself, was shouting incoherent instructions to me.

"Get the lamb! Gerry, get the lamb! They'll kill it," she squeaked, covered with lemonade and bits of chocolate cake.

"Black son of the devil! Illegitimate offspring of a witch! My shoes! Leave my shoes! I will kill you . . .

destroy you!" panted the old Turk, slashing away at Roger.

"Ayii! Ayii! Ayii! His shoes! His shoes!" screamed the women in a chorus, immobile on their cushions.

With difficulty and without getting a stab wound myself, I managed to tear the ravening Roger off the old Turk's pompoms and get him and Widdle and Puke out onto the verandah. Then I opened the sliding doors and shut the lamb in the dining room as a temporary measure while I soothed the old Turk's wounded feelings. Mother, smiling nervously and nodding vigorously at everything I said, although she did not understand it, was making an attempt to clean herself up, but this was rather ineffectual as the chocolate cake had been one of her larger and more glutinous creations, oozing cream, and she had put her elbow into the exact center of it as she fell backwards. At length I managed to soothe the old man, and while Mother went up to change her dress I dished out brandy to the Turk and his three wives. My helpings were liberal, so by the time Mother came back, faint hiccups were coming from behind at least one of the veils and the Turk's nose had turned fiery red.

"Your sister is . . . how shall I tell you? . . . magical . . . God-given. Never have I seen a girl like her," he said, holding out his glass eagerly. "I, who, as you see, have three wives, I have never seen a girl like your sister."

"What's he saying?" asked Mother, eyeing his dagger nervously. I repeated what the Turk had said.

"Disgusting old man," said Mother. "Really, Margo should be more careful."

The Turk drained his glass and held it out again, beaming convivially at us.

"Your maid here," he said, jerking a thumb at Mother, "she is a little bit soft, huh? She doesn't speak Greek."

"What does he say?" asked Mother.

Dutifully, I translated.

"Impertinent man!" said Mother indignantly. "Really, I could smack Margo. Tell him who I am, Gerry."

I told the Turk, and the effect on him was more than Mother could have wished. With a roar, he leapt to his feet, rushed across to her, seized her hands and covered them with kisses. Then, still holding her hands in a vicelike grip, he peered into her face, his moustache trembling.

"The mother," he intoned, "the mother of my Almond Blossom."

"What's he say?" asked Mother tremulously.

But before I could translate, the Turk had barked out an order to his wives, who showed their first sign of animation. They leapt from their cushions, rushed to Mother, lifted their yashmaks, and kissed her hands with every symptom of veneration.

"I do wish they wouldn't keep kissing me," gasped Mother. "Gerry, tell them it's quite unnecessary."

But the Turk, having got his wives reestablished on their cushions, turned once again to Mother. He threw a powerful arm around her shoulders, making her squeak, and threw out his other arm oratorically." Never did I think," he boomed, peering into Mother's face, "never did I think that I should have the honor of meeting the mother of my Almond Blossom."

"What's he saying?" asked Mother agitatedly, trapped in the Turk's bearlike hug.

I translated faithfully.

"Almond Blossom? What's he talking about? The man's mad," she said.

I explained that the Turk was apparently greatly enamored of Margo and that this was his name for her.

This confirmed Mother's worst fears about the Turk's intentions. "Almond Blossom, indeed!" she said indignantly. "Just wait until she gets back – I'll give her Almond Blossom!"

Just at that moment, cool and fresh from a swim, Margo herself appeared in a very revealing bathing costume.

"Ooooh!" she screamed delightedly. "Mustapha! And Lena, and Maria, and Telina! How lovely!"

The Turk rushed across to her and kissed her hands reverently while his wives clustered round making muffled noises indicative of pleasure.

"Mother, this is Mustapha," said Margo, glowing.

"We have already met," said Mother grimly, "and he's ruined my new dress, or rather his lamb has. Go and put some clothes on."

"His lamb?" asked Margo, bewildered. "What lamb?"

"The lamb he brought for his Almond Blossom, as he calls you," said Mother accusingly.

"Oh, it's just a nickname," said Margo, coloring. "He doesn't mean any harm."

"I know what these old men are," said Mother ominously. "Really, Margo, you should know better."

The old Turk was listening to this exchange with quick glances from his bright eyes and a beatific smile on his face. However, I could see that my powers of translation would be stretched to their limit if Mother and Margo started arguing, so I opened the sliding doors and let the lamb in. He came in pertly, prance-footed, black and curly as a storm cloud.

"How dare you!" said Margo. "How dare you insult my friends? He's not a dirty old man; he's one of the cleanest old men I know."

"I don't care whether he's clean or not," said Mother,

coming to the end of her patience. "He can't stay here with all his . . . his . . . women. I'm not cooking for a harem."

"It is wonderful to hear the mother and daughter talk together," the Turk confided to me. "It's like the sound of sheep bells."

"You're beastly," said Margo, "you're beastly! You don't want me to have any friends. You're narrow-minded and suburban!"

"You can't call it suburban to object to three wives," said Mother indignantly.

"It reminds me," said the Turk, his eyes moist, "of the singing of the nightingales in my valley."

"He can't help it if he's a Turk," shrilled Margo. "He can't help it if he's got to have three wives."

"Any man can avoid having three wives if he puts his mind to it," said Mother firmly.

"I expect," said the Turk confidingly, "Almond Blossom is telling her mother what a happy time we had in my valley, huh?"

"You always try to repress me," said Margo. "Everything I do is wrong."

"The trouble is, I give you too much license. I let you go away for a few days and you come back with this . . . this . . . old roué and his dancing girls," said Mother.

"There you are, that's what I mean – you repress me," said Margo triumphantly. "Now you expect me to have a license for a Turk."

"How I would like to take them back to my village," said the Turk, gazing at them fondly. "Such wonderful time we would have . . . dancing, singing, wine. . . ."

The lamb seemed disappointed that no one was taking any notice of him; he had gamboled a little, decorated the floor, and done two nicely executed pirouettes, but he felt that no one was paying him the attention he deserved, so he put down his head and charged Mother. It was a beautifully executed charge. I could speak with some authority, for during my expeditions through the surrounding olive groves I had frequently met with eager and audacious young rams and fought them matador-fashion, using my shirt as a cloak, to our mutual satisfaction. So I considered myself to be something of an expert and, while deploring the result, I had to confess that the charge was excellent, well thought out, and with the full power of the ram's wiry body and bony head landing with precision on the back of Mother's knees. Mother was projected onto our extremely uncomfortable horsehair sofa as if propelled by a cannon, and she lay there gasping. The Turk, horrified at what his gift had done to her, leapt in front of her, arms outstretched, to protect her from further attack, which seemed imminent, for the ram, pleased with itself, had retreated to a corner of the room and was prancing and bucking rather in the manner of a boxer limbering up in his corner of the ring.

"Mother, Mother, are you all right?" screamed Margo.

Mother was too breathless to answer her.

"Ah-ha! You see, he has spirit like me, Almond Blossom," cried the Turk. "Come on then, my brave one, come on!"

The ram accepted the invitation with a speed and suddenness that took the Turk by surprise. It moved across the room in a black blur, its feet machine-gunning on the scrubbed boards, hit the Turk on his shins with a crack, and precipitated him onto the sofa with Mother, where he lay uttering loud cries of rage and pain. I had been charged in the shins like that, so I could sympathize.

The Turk's three wives, aghast at their master's downfall, were standing immobile, uttering noises like three minarets at sundown. It was into this interesting situation that Larry and Leslie intruded. They stood riveted in the doorway, drinking in the scene with unbelieving eyes. There was I pursuing a recalcitrant lamb round the room, Margo comforting three ululating ladies in veils, and Mother apparently rolling around on the sofa with an elderly Turk.

"Mother, don't you think you're getting a little old for this sort of thing?" Larry asked with interest.

"By Jove, look at that marvelous dagger," said Leslie, eyeing the still writhing Turk.

"Don't be stupid, Larry," said Mother angrily, massaging the backs of her legs. "It's all Margo's Turk's fault."

"You can't trust Turks," said Leslie, still eyeing the dagger. "Spiro says so."

"But what are you doing rolling about with a Turk at this hour?" Larry inquired. "Practicing to be Lady Hester Stanhope?"

"Now, Larry, I've had quite enough this afternoon. Stop making me angry. The sooner this man is out of here, the sooner I'll be pleased," said Mother. "Kindly ask him to go."

"You can't, you can't. He's my Turk," squeaked Margo tearfully. "You can't treat my Turk like that."

"I'm going upstairs to put some witch hazel on my bruises," said Mother, hobbling towards the door, "and I want that man out of here by the time I come down."

By the time she had returned, both Larry and Leslie had struck up a firm friendship with the Turk, and to Mother's annoyance he and his wives stayed on for several hours, imbibing gallons of sweet tea and biscuits before we could finally manage to get them into a *caraccino* and back to town.

"Well, thank heaven *that's* over," said Mother, limping towards the dining room for our evening meal. "At least they're not staying here, and that's one mercy. But really, Margo, you should be careful whom you invite."

"I'm sick of the way you criticize my friends," said Margo. "He was a perfectly ordinary, harmless Turk."

"He would have made a charming son-in-law, don't you think?" asked Larry. "Margo could have called the first son Ali Baba and the daughter Sesame."

"Don't joke like that, Larry dear," said Mother.

"I'm not joking," said Larry. "The old boy told me his wives were getting a bit long in the tooth and that he rather fancied Margo as number four."

"Larry! He didn't! Disgusting old brute," said Mother. "It's a good thing he didn't say that to me. I'd have given him a piece of my mind. What did *you* say?"

"He was rather put off when I told him what Margo's dowry was," said Larry.

"Dowry? What dowry?" asked Mother, mystified.

"Eleven unweaned puppies," explained Larry.

Ghosts and Spiders

Take heed o' the foul fiend.
SHAKESPEARE, *King Lear*

THROUGHOUT THE YEAR Thursday was, as far as I was concerned, the most important day of the week, for that was the day that Theodore visited us. Sometimes it would be a long family day we planned – a drive down south and a picnic on a remote beach, or something similar; but normally Theo and I would set off alone on one of our excursions, as he insisted on calling them. Bedecked with our collecting equipment and bags, nets, bottles and test tubes and accompanied by the dogs, we would set out to explore the island in much the same spirit of adventure that filled the bosoms of Victorian explorers who ventured into darkest Africa.

But not many of the Victorian explorers had had the benefit of such a companion as Theodore; as a handy encyclopedia to take along on a trip, he could not be bettered. To me he was as omniscient as a god, but much nicer since he was tangible. It was not only his incredible erudition that astonished everyone who met him, but also his incredible modesty. I remember how we would sit on the verandah, surrounded by the remnants of one of Mother's sumptuous teas, listening to the tired cicadas singing the evening in and plying Theodore with questions. As he sat meticulously dressed in his tweed suit, his blonde hair and beard immaculate, his eyes would sparkle with interest as each new subject was introduced.

"Theodore," Larry would ask, "there's a painting up in the monastery at Paliocastritsa that the monks say was done by Panioti Dokseras. D'you think it is?"

"Well," Theodore would say cautiously, "I'm afraid it's a subject about which I know very little. But I believe I'm right in saying that it's more likely to be the work of Tsadzanis . . . er . . . he did that most interesting little picture . . . in the Patera Monastery . . . you know, the one on the upper road leading north of Corfu. Now, he, of course. . . ."

During the next half hour he would give an all-embracing and succinct lecture on the history of painting in the Ionian Islands since about 1242 and then end by saying, "But if you want an *expert* opinion, there's Doctor Paramythiotis, who'd give you much more information than I can."

It was small wonder that we treated him like an oracle. The phrase "Theo says" set the seal of authenticity on whatever item of information one was going to vouchsafe; it was the touchstone for getting Mother's agreement to anything from the advisability of living entirely on fruit to get slim to the complete innocuousness of keeping scorpions in one's bedroom. Theodore was everything to everyone. With Mother he could discuss plants, particularly herbs and recipes, while keeping her supplied with reading matter from his capacious library of detective novels. With Margo he could discuss diets, exercises and the various unguents supposed to have a miraculous effect on spots, pimples and acne. He could keep pace effortlessly with any idea that entered the mercurial mind of my brother Larry, from Freud to peasant belief in vampires. And Leslie he could enlighten on the history of firearms in Greece or the winter habits of the hare. As far as I was concerned, with my hungry, questing and ignorant mind, Theodore represented a fountain of knowledge on every subject, from which I drank greedily. On Thursday Theodore would generally arrive at about ten, sitting sedately in the back of the horse-drawn cab, silver homburg on his head, his collecting box on his knees, his walking stick with its little gauze net on the end by his side. I, who had been up since six peering down through the olive groves to see if he was coming, would by now have decided in despair that he had forgotten what day it was or that he had fallen down and broken his leg or that some other awful catastrophe had overtaken him. My relief at seeing him, grave, sedate and intact in the back of the cab, would be considerable. The sun, up until then suffering from an eclipse, would start to shine again. Having shaken me by the hand courteously, Theodore would pay the cabman, remind him to return at the appropriate hour in the evening, and then, hoisting his collecting bag onto his shoulders, he would contemplate the ground, rising and falling on his well-polished boots.

"I think . . . er . . . you know," he would say, "we might investigate those little ponds near . . . er . . . Kontokali. That is to say, unless there is somewhere else . . . er . . . you know . . . that you would prefer to go."

I would say happily that the little ponds near Kontokali would suit me fine.

"Good," Theodore would say. "One of the reasons I particularly want to go . . . er . . . that way . . . is because the path takes us past a very good ditch . . . er . . . you know . . . that is to say, a ditch in which I have found a number of rewarding specimens."

Talking happily, we would set out, and the dogs,

tongues lolling, tails wagging, would leave the shade of the tangerine trees and follow us. Presently a panting Lugaretzia would catch up with us, carrying the bag containing our lunch, which we had both forgotten.

We would make our way through the olive groves, chattering together, stopping periodically to examine a flower or a tree, a bird or a caterpillar; everything was grist to our mill, and Theodore knew something about everything.

"No, I don't know of any way you could preserve mushrooms for your collection; whatever you use, they would . . . um . . . er . . . you know . . . shrivel up. The best way would be to draw or paint them, or, perhaps, you know, *photograph* them. You could collect the spore patterns, though, and they are remarkably pretty. What? Well, you remove the cup of the . . . er . . . you know . . . the mushroom or toadstool and place it on a white card. The fungus must be ripe, of course, or it won't drop its spores. After a time, you remove the cap carefully from the card . . . that is to say, you take care not to smudge the spores, and you will find an attractive . . . er . . . sort of pattern is left."

The dogs would fan out ahead of us, cocking their legs, snuffling in the dark holes that honeycombed the great, ancient olive trees, and dashing off in noisy and futile pursuit of the swallows that skimmed only millimeters above the ground down the long meandering avenues of trees. Presently we would reach more open country, where the olive groves would give way to small fields of fruit trees and maize or vineyards.

"Aha!" Theodore would say, stopping by a weedy, waterfilled ditch and peering into it, his eyes gleaming, his beard bristling with enthusiasm. "Now here's something interesting. There, do you see? Just by the end of my stick."

I would strain my eyes but see nothing. Theodore, attaching his net to the end of his walking stick, would make a neat dipping motion, like a man taking a fly out of his soup, and would then haul in the net.

"There, you see? It's the egg sac of the *Hydrophilus piceus* . . . er . . . that is to say, the great silver water beetle. It's the female, as you know, that spins . . . er . . . makes this sac. It may have up to fifty eggs in it; the curious thing is – just a minute, while I get my forceps. Um . . . there . . . you see? Now, this . . . um . . . you might say chimney, though perhaps mast would be better, is filled with air so the whole thing is rather like a little boat which can't capsize. The . . . er . . . air-filled mast prevents it. Yes, if you put it in your aquarium, it should hatch out, though I must warn you that the larvae are very . . . er . . . you know . . . very *fierce* and will probably eat your other specimens. Let's see if we can catch an adult."

Patient as any wading bird, Theodore would pace the edge of the ditch, dipping his net in at intervals and sweeping it to and fro.

"Aha! Success!" he would exclaim at last and carefully place a large black beetle, legs thrashing indignantly, into my eager hands.

I would admire the strong, ribbed wing cases, the bristly legs, the whole body with a faint olive-green sheen.

"It's a rather slow swimmer compared with the other . . . er . . . you know.. aquatic beetles, and it has a very curious method of swimming. Um . . . um . . . instead of using the legs together, like any other aquatic species, it uses them alternately. It gives it a . . . you know . . . very jerky appearance."

The dogs, on these occasions, were somewhat of a mixed blessing. Sometimes they would distract us by rushing into a peasant's farmyard and attacking all his chickens, the ensuing altercation with the chicken owner wasting at least half an hour; and at other times they would be quite useful, surrounding a snake so it could not escape and barking prodigiously until we came to investigate. For me, at any rate, they were comforting to have around – Roger, like a stocky, unclipped black lamb; Widdle, elegant in his silky coat of fox-red and black; and Puke, looking like a miniature liver and white spotted bull terrier. They liked accompanying us, though occasionally they would get bored if we stopped for too long, but generally they lay patiently in the shade, pink tongues flicking, tails wagging amicably whenever they caught our eye. Sometimes they could be useful. It was Roger, for example, that first introduced me to one of the most beautiful spiders in the world, with the elegant-sounding name of *Eresus niger*.

We had walked a considerable distance, and at noon, when the sun was at its hottest, we decided to stop and eat our picnic in the shade. We sat down at the edge of an olive grove and started feasting on sandwiches and ginger beer. Normally, when Theodore and I had our meal, the dogs would sit around panting and gazing at us imploringly, since they were always of the opinion that our food was superior to their food in some way, and so having finished their rations they would try and obtain largesse from us, using all the wiles of an Asiatic beggar. On this particular occasion, Widdle and Puke rolled their eyes, panted and gasped, and tried by every means possible to show us that they were at death's door from starvation. But, unusually, Roger did not join in. Instead he was sitting out in the sunshine in front of a patch of brambles watching something with great intentness.

I went over to see what was intriguing him to such an extent that he was ignoring my sandwich crusts. At first I could not see what it was, and then suddenly I saw something so startlingly beautiful that I could hardly believe my eyes. It was a tiny spider, the size of a pea, and at first glance it looked like an animated ruby or a moving drop of blood. Uttering a whoop of joyous enthusiasm, I rushed to my collecting bag and got a glass-topped pillbox in which to catch this brilliant creature. He was not easily caught, however, for he could take prodigious jumps for his size, and I had to pursue him round and round the bramble bush for some time before I had him safely locked in my pillbox. Triumphantly I carried this gorgeous spider over to Theodore.

"Aha!" said Theodore, taking a swig of ginger beer before producing his magnifying glass the better to examine my capture. "Yes an *Eresus niger* . . . um . . . yes . . . this is of course the male, such a pretty creature. The female is . . . er . . . you know . . . all black, but the male is very brightly colored."

On close examination through a magnifying glass, the spider turned out to be even more beautiful than I had thought. His forequarters, or cephalothorax, were velvety black with little specks of scarlet at the edges. His rather stocky legs were ringed with white bands, so he looked ridiculously as though he were wearing striped pants. But it was his abdomen that was really eye-catching. It was vivid huntsman's red, marked with three round black spots rimmed with white hairs. He really was the most spectacular spider that I had ever seen, and I was determined to try and get him a mate and see if I could breed them. I subjected the bramble bush and the terrain around it to a minute scrutiny, but with no success. Theodore explained to me that the female spider digs a burrow about three inches long and lines it with tough silk. "You can distinguish it from other spiders' burrows," he said, "because the silk at one point is protruded like an apron, and this forms a sort of roof over the mouth of the tunnel. Moreover, the outside is covered with bits and pieces of the female spider's past meals, in the shape of grasshopper legs and wing cases and the remains of beetles."

Armed with this knowledge, I went the following day and combed the entire area round the bramble bush once again. But after spending the whole afternoon at it, I still did not meet with success. Irritably I started on my way home to tea. I took a short cut that led me over some small hills covered with the giant Mediterranean heather, which seemed to flourish in this sandy and rather desiccated terrain. It was the sort of wild, dry country favored by ant lions, fritillaries and other sun-loving butterflies, lizards and snakes. As I walked along, I suddenly came upon the ancient skull of a sheep, and in one of its empty eye sockets a praying mantis had laid its curious egg case, which to me always looked like an oval pudding of some sort, made out of ribbed sponge cake. I was squatting down examining this and wondering whether to take it to the villa and add it to my collection when I suddenly saw the burrow of a female spider, just as Theodore had described it.

I pulled out my knife and with great care excavated a large wedge of soil, which when levered out, contained not only the spider but her burrow as well. Delighted with my success, I placed it carefully in my collecting bag and hurried back to the villa. I had already got the male installed in a small aquarium, but I felt that the female was worthy of better things. So I unceremoniously evicted two frogs and a baby tortoise from my largest aquarium and made it ready for her. When I had decorated it with bits of heather and interesting branches of moss, I put the wedge of earth containing her and her nest on the bottom and left her to recover from this sudden and unexpected house move.

Three days later I introduced the male. At first it was very dull, because he did nothing more romantic than rush about like an animated hot ember, trying to catch the various insects I had put in the aquarium as provender. But then, early one morning when I went to look, I discovered that he had found the lair of the female. He was walking to and fro around it, in a curious jerky fashion, his striped legs stiff, his body trembling with what one could only conclude was passion. He strutted about in this great state of excitement for a minute or so and then approached the burrow and disappeared under the roof. To my annoyance I could no longer observe him, but I presumed that he must be mating with the female. He was in the burrow for about an hour, and then he emerged jauntily and continued his carefree pursuit of the bluebottles and grasshoppers that I had provided for him. However, I removed him to another aquarium as a precautionary measure, since I knew that in some species the female had cannibalistic habits and was not averse to making a light snack of her husband.

The rest of the drama I could not witness in detail, but I saw bits of it, as it were. The female eventually laid a bunch of eggs, which she carefully encapsulated in a web. This balloon of eggs she stored down in her tunnel, but she brought it up each day to hang under the roof. Whether she did this so that the eggs could get more heat from the sun or whether it was to allow them access to more fresh air, I was not certain. The egg case was disguised by means of small morsels of beetle and grasshopper remains attached to the outside.

As the days passed she proceeded to add to the roof over the tunnel, and then finally she constructed a silken room aboveground. I watched this architectural achievement for a considerable time and then, as I could see nothing, I grew impatient. With the aid of a scalpel and a long darning needle I carefully opened up the silken room. To my astonishment I found that it was surrounded by cells in which all of the young spiders sat, while in the central hall, as it were, lay the corpse of their mother. It was rather a macabre, yet touching sight, the babies all sitting round the mortal remains of their mother in a sort of spiders' wake. When the babies hatched, however, I was forced to let them all go, since providing food for some eighty minute spiders raised a problem in catering which even I, enthusiastic though I was, could not solve.

Among the numerous friends that Larry saw fit to inflict on us were a strange pair of painters called Lumis Bean and Harry Bunny. They were both American and deeply devoted to each other, so much so that within twenty-four hours they were known privately to the family as Lumy Lover and Harry Honey. They were young and very good-looking, with the fluid boneless grace of movement that you expect from colored people but rarely get in a European. They wore perhaps a shade too many gold bangles and a soupçon too much scent and hair cream, but they were nice and, what was rather unusual in the painters who came to stay, hardworking. Like so many Americans, they were possessed of a charming naïveté and earnestness, but these qualities, as far as Leslie was concerned at any rate, made them the ideal subjects for practical jokes. I used to participate in these and then relate the results to Theodore, who got as much innocent pleasure from them as Leslie and I did. In fact, I sometimes got the feeling that Theodore looked forward to my report on the jokes with more interest than he did to my weekly report on the progress of my menagerie.

Leslie had rather a genius for practical jokes, and the childlike innocence of our two guests inspired him to soar to new heights. Shortly after their arrival he got them to congratulate Spiro most prettily on his final success in taking out Turkish naturalization papers. Spiro, who, like most Greeks, considered the Turks to be slightly more malevolent than Satan himself and who had spent several years fighting them, exploded like a volcano. Fortunately Mother was near at hand and moved swiftly between the white-faced, protesting, bewildered Lumy and Harry and Spiro's barrel-shaped, muscular bulk. She looked not unlike a diminutive Victorian missionary facing a charging rhino.

"Gollys, Mrs. Durrells," Spiro roared, his gargoyle features purple with rage, his hamlike hands clenched, "lets me pokes them one."

"Now, now, Spiro," said Mother, "I'm sure it's all a mistake. I'm sure there's an explanation."

"They calls me a bastard Turks!" roared Spiro. "I'm Greeks. I'm no bastard Turks."

"Of course you're not," said Mother soothingly. "I'm sure it was just a mistake."

"Mistakes!" bellowed Spiro, his plurals coming thick and fast with rage. "Mistakes! I'm nots goings to be called a bastards Turks by these bloody fairies, if you'll excuses my language, Mrs. Durrells."

It was some considerable time before Mother could calm Spiro and get a coherent story out of the terrified Lumy Lover and Harry Honey, and the whole episode gave her a severe headache. She was very cross with Leslie.

Some time later Mother had to move them out of the bedroom we had given them, because it was going to be decorated. She put them temporarily into one of our large, gloomy attics. This gave Leslie the opportunity of telling them the story of the headless bell ringer of Kontokali, who died in the attic. He was the fiend who in 1604, or thereabouts, was official executioner and torturer in Corfu. First he would torture his victims and then he would ring his bell before they were finally beheaded. Getting slightly fed up with him, the villagers of Kontokali broke into the villa one night and beheaded the bell ringer, and now, as a prelude to seeing his ghost, headless and with a gory stump, you would hear him frantically ringing his bell.

Having convinced our earnest couple of the authenticity of this fable by getting it vouched for by Theodore, Leslie borrowed fifty-two alarm clocks from a friendly clockmaker in town, prized up two floorboards in the attic, and placed the clocks, all set to go off at three in the morning, carefully between the joists.

The effect of fifty-two alarm clocks all going off simultaneously was most gratifying. Not only did Lumy and Harry vacate the attic with all speed, uttering cries of terror, but in their haste they tripped each other up and, clasped in each other's arms, fell heavily down the attic stairs. The resulting turmoil woke the whole house, and it was some time before we could convince Lumy and Harry that it was a joke and soothe them with brandy. Mother, together with our guests, had a severe headache all next day and would hardly talk to Leslie at all.

The affair of the invisible flamingoes came about one day quite casually as we were sitting having tea on the verandah. Theodore had asked our pair of Americans how their work was progressing.

"Darling Theo," said Harry Honey, "we're getting on divinely, simply divinely, aren't we, lover?"

"We sure are," said Lumy Lover, "we sure are. The light here is fantastic, simply fantastic. It's as though the sun were closer to the earth somehow, you know."

"It sure does seem that way," Harry Honey agreed. "It seems just like, as Lumy says, the sun is right down low, beaming straight at little old us."

"I said that to you this morning, Harry honey, didn't I?" said Lumy Lover.

"You did, Lumy, you did. Right up there by that little barn, do you remember, you said to me – "

"Have another cup of tea," Mother interrupted, for she knew from experience that these postmortems to prove the togetherness of the two could go on indefinitely.

The conversation drifted on into the realms of art, and I scarcely listened until suddenly my attention was riveted by Lumy Lover saying, "Flamingoes! Ooh, Harry honey, flamingoes! My favorite birds. Where, Les, where?"

"Oh, over there," said Leslie, giving a wave that embraced Corfu, Albania and the better half of Greece. "Great flocks of them."

Theodore, I could see, was holding his breath, as was I, in case Mother, Margo or Larry should say anything to upset this outrageous lie.

"Flamingoes?" said Mother interestedly. "I didn't know there were any flamingoes here."

"Yes," said Leslie solemnly, "hundreds of them."

"Did you know there were flamingoes, Theodore?" asked Mother.

"I . . . er . . . you know . . . caught a glimpse of them down on Lake Halikiopoulos," said Theodore, not deviating from the truth but not adding that this had been three years previously and the only time that flamingoes had ever visited Corfu. I had a handful of pink feathers to commemorate it.

"Jee-hovah!" said Lumy Lover. "Could we catch a glimpse of them, Les dear? D'you suppose we could sneak up on them?"

"Sure," said Leslie airily, "easiest thing in the world. They migrate over the same route every day."

The following morning Leslie came into my room carrying what looked like a strange form of trumpet made out of a cow's horn. I asked him what it was and he grinned.

"It's a flamingo decoy," he said with satisfaction.

I was deeply interested and said I had never heard of a flamingo decoy.

"Neither have I," Leslie admitted. "It's an old cow's horn powder container, for muzzleloaders, you know. But the end's broken off so you can blow on it."

By way of illustration, he raised the pointed end of the cow's horn to his lips and blew, and the horn produced a long, sonorous sound somewhere between a foghorn and a raspberry, with very vibrant overtones. I listened critically and said that it did not sound a bit like a flamingo.

"Yes, but I bet Lumy Lover and Harry Honey don't know that," said Leslie. "Now all I need is to borrow your flamingo feathers."

I was somewhat reluctant to part with such rare specimens from my collection until Leslie explained why he wanted them and said that they would come to no harm.

At ten o'clock Lumy and Harry appeared, having been

dressed by Leslie for flamingo hunting. Each wore a large straw hat and gumboots, for, as Leslie explained, we might have to follow the flamingoes into the swamps. Lumy and Harry were flushed and excited at the prospect of this adventure, and their enthusiasm when Leslie demonstrated the flamingo decoy knew no bounds. They blew such enthusiastic blasts on it that the dogs went mad and howled and barked, and Larry, furious, leaned out of his bedroom window and said that if we were all going to carry on like a meet of the bloody Quorn hunt he was going to move.

"And you're old enough to know better!" was his parting shot as he slammed the window. This was addressed to Mother, who had just joined us to see what the noise was all about.

We eventually got our bold hunters into the field and walked them about two miles, by which time their enthusiasm for flamingo hunting was on the wane. Then we scrambled them up to the top of an almost inaccessible hillock, stationed them inside a bramble bush as a disguise, and told them to keep calling to attract the flamingoes. For half an hour they blew on the horn in turns with great dedication, but then their wind started to give out, and towards the end the noise they were making was beginning to sound more like the despairing cry of a mortally wounded bull elephant than anything in the bird line.

Then it was my turn. Panting and excited, I rushed up the hillock and told our hunters that their efforts had not been in vain. The flamingoes had indeed responded, but unfortunately they had settled in a valley below a hill half a mile east. If they hurried there, Leslie was waiting. I will say that I was lost in admiration at their American tenacity. Thumping along in their ill-fitting gumboots, they galloped off to the distant hill, pausing periodically as per my instructions to blow gaspingly on the flamingo decoy. When, in an ocean of sweat, they reached the top of the hill, they found Leslie. He said that if they remained there and continued to blow on the decoy, he would make his way around the valley and drive the flamingoes up to them. He gave them his gun and game bag so that, as he explained, he could stalk more easily, and then he faded away.

It was at this point that our favorite policeman, Filimona Kontakosa, entered the act. Filimona was without doubt the fattest and most somnambulistic of all the Corfu policemen; he had been in the force for thirty-odd years and owed his lack of promotion to the fact that he had never made an arrest. He had explained to us at great length that he was, in fact, physically incapable of making an arrest; the mere thought of being harsh to a criminal would fill his pansy-dark eyes with tears, and on feast days, at the slightest sign of altercation among the wine-happy villagers, you could see him waddling resolutely in the opposite direction. He preferred to lead a gentle, benign life, and every fortnight or so he would pay us a visit to admire Leslie's gun collection (for which we had no permits) and bring gifts of smuggled tobacco to Larry, flowers to Mother and Margo, and sugared almonds to me. He had, in his youth, been a deckhand on a cargo boat and had acquired a tenuous command of the English language, and this, combined with the fact that all Corfiotes adore practical jokes, made him perfect for our purposes. He rose to the occasion magnificently.

He waddled to the top of the hill, resplendent in his uniform, every kilo of him looking the personification of law and order and a credit to the force. He found our hunters blowing in a desultory fashion on their decoy.

Benignly he asked them what they were doing. Responding to kindness like two puppies, Lumy Lover and Harry Honey were only too delighted to compliment Filimona on his truncated English and explain matters to him. To the Americans' consternation, he suddenly changed from being a kindly, twinkling, fat policeman to being the cold, brutal representative of officialdom.

"You no know flamongoes you no shoot?" he snapped at them. "Is forbidden shoot flamongoes!"

"But, darling, we're not shooting them," said Lumy Lover falteringly. "We only want to *see* them."

"Yes. Gee, you got it all wrong," said Harry Honey ingratiatingly. "We don' wanna shoot the little fellas; we just wanna see 'em. No shoot, see?"

"If you no shoot, why you have gun?" asked Filimona.

"Oh, that," said Lumy Lover, reddening. "That belongs to a friend of ours . . . er . . . *amigo* . . . savvy?"

"Yeah, yeah," said Harry Honey, "friend of ours, Les Durrell. Maybe you know him? He's well known around these parts."

Filimona stared at them coldly and implacably.

"I no know this friend," he said at last. "Please to open bag."

"Well, now, steady on, see here," protested Lumy Lover. "This isn't *our* bag, officer."

"No, no," said Harry Honey, "it belongs to this friend of ours, Durrell."

"You have gun. You have bag," Filimona pointed out. "Please to open bag."

"Well, I must say, I think you're exceeding your duty just a tiny bit, officer, I really do," said Lumy Lover, while Harry Honey nodded eager assent. "But if it'll make you feel any easier, well then, I don't suppose there's much harm in letting you have a little peek."

He wrestled briefly with the straps of the bag, opened it and handed it to Filimona. The policeman peered into it, gave a triumphant grunt and pulled from the interior the plucked and headless body of a chicken, to which were adhering numerous bright pink feathers. Both stalwart flamingo hunters went white with emotion.

"But see here now . . . er . . . wait a moment," Lumy Lover began, and his voice trailed away before Filimona's accusing look.

"Is forbidden shoot flamongo, I tell you," said Filimona. "I arrest you both."

He herded them, alarmed and protesting, down to the village police station and kept them there for several hours, during which they nearly went mad writing out statements and getting so muddled through nerves and frustration that they kept contradicting each other's stories. To add to their alarm, Leslie and I had assembled a crowd of our village friends, who shouted and roared in the terrifying way Greeks have, periodically shouting, "Flamongo!" and throwing the odd stone at the police station.

Eventually Filimona allowed his captives to send a note to Larry, who stormed down into the village, told Filimona it would be more to the point if he caught some evildoers rather than indulging in practical jokes, and brought our two flamingo hunters back to the bosom of the family.

"This has got to stop!" said Larry angrily. "I will not have my guests subjected to ill-bred japes perpetrated by a pair of half-witted brothers."

I must say Lumy Lover and Harry Honey were wonderful.

"Don't be angry, Larry darling," said Lumy Lover, "it's just high spirits. It's just as much our fault as Les's."

"Yes," said Harry Honey, "Lumy's right. It's our fault for being so gullible, silly old us."

To show that there was no ill-feeling, they went down to the town and brought back a crate of champagne to have a party. They even went down to the village themselves to fetch Filimona up to the house for it. They sat on the terrace, one on each side of the policeman, toasting him coyly with champagne while Filimona, in a surprisingly good tenor, sang love songs that brightened his great dark eyes with tears.

"You know," Lumy Lover confided to Larry at the height of the party, "he'd be really very good-looking if he went on a diet. But don't tell Harry I said so, darling, will you?"

The Merriment of Friendship

The sound of the cornet, flute, harp, sackbut, psaltery, dulcimer, and all kinds of musick.

DANIEL 3:5

IT WAS TOWARDS the end of summer that we held what came to be known as our Indian party. Our parties, whether carefully planned or burgeoning on the spur of the moment out of nothing, were always interesting affairs, since things seldom went exactly as we planned them. In those days, living as we did in the country, without the dubious benefits of radio or television, we had to rely on such primitive forms of amusement as books, quarreling, parties, and the laughter of our friends, so naturally parties – particularly the more flamboyant ones – became red-letter days, preceded by endless preparations, and even when they were successfully over, they provided days of delightfully acrimonious argument as to how they could have been better stage-managed.

We had had a fairly tranquil patch for a month or so; we had not had a party, and no one had turned up to stay, so Mother had relaxed and become very benign. We were sitting on the verandah one morning reading our mail when the Indian party was hatched. In her mail Mother had just received a mammoth cookery book entitled *A Million Mouthwatering Oriental Recipes*, lavishly illustrated with color reproductions so lurid and glossy that you felt you could eat them. Mother was enchanted with it and kept reading bits aloud to us.

"Madras Marvels!" she exclaimed delightedly. "Oh, they're lovely. I remember them. They were a favorite of your father's when we lived in Darjeeling. And look! Konsarmer's Delights! I've been looking for a recipe for them for *years*. They're simply delicious, but so rich."

"If they're anything like the illustrations," said Larry, "you'd have to live on a diet of bicarbonate of soda for the next twenty years after you ate one."

"Don't be silly, dear. The ingredients are absolutely pure – four pounds of butter, sixteen eggs, eight pints of cream, the flesh of ten young coconuts – "

"God!" said Larry. "It sounds like a breakfast for a Strasbourg goose."

"I'm sure you'll like them dear," said Mother. "Your father was very fond of them."

"Well, I'm supposed to be on a diet," said Margo. "You can't go forcing me to have stuff like that."

"Nobody's forcing you, dear," said Mother. "You can always say no."

"Well, you know I can't say no, so that's forcing," said Margo.

"Go and eat in another room," suggested Leslie, flipping through the pages of a gun catalogue, "if you haven't got the willpower to say no."

"But I *have* got the willpower to say no," said Margo indignantly. "I just can't say no when Mother offers it to me."

"Jeejee sends his salaams," said Larry, looking up from the letter he was perusing. "He says he's coming back here for his birthday."

"His birthday!" exclaimed Margo. "Ooh, good, I'm glad he remembered."

"Such a *nice* boy," said Mother. "When's he coming?"

"As soon as he gets out of hospital," said Larry.

"Hospital? Is he ill?"

"No, he's just having trouble with his levitation; he's got a busted leg. He says his birthday's on the sixteenth so he'll try and make it by the fifteenth."

"I *am* glad," said Mother. "I grew very fond of Jeejee and I'm sure he'll love this book."

"I know, let's give him a huge birthday party," said Margo excitedly. "You know, a really *huge* party."

"That's a good idea," said Leslie. "We haven't had a decent party for ages."

"And I could make him some of the recipes out of this book," said Mother, obviously intrigued by the thought.

"An oriental feast," said Larry. "Tell everyone to come in turbans, with jewels in their navels."

"No, I think that's going too far," said Mother. "No, let's just have a nice, quiet, little – "

"You can't just have a nice, quiet little party for Jeejee," said Leslie. "Not after you told him you always traveled with four hundred elephants. He expects something a bit spectacular."

"It wasn't four hundred elephants, dear. I only said we went *camping* with elephants," said Mother. "You children do exaggerate. And anyway, we can't produce elephants here; he wouldn't expect it."

"No, but you've got to put on some sort of show," said Leslie.

"I'll do all the decorations," said Margo. "Everything will be oriental. I'll borrow Mrs. Papadrouya's Burmese screens, and there are the ostrich feathers that Lena's got . . . "

"We've still got a wild boar and some duck and stuff left in the cold room in town," said Leslie. "Better use it up."

"I'll borrow Countess Lefraki's piano," said Larry.

"Now, look, all of you . . . stop it," said Mother, alarmed. "It's not a durbar we're having, just a birthday party."

"Nonsense, Mother, it'll do us good to let off a little steam," said Larry indulgently.

"Yes, in for a penny, in for a pound," said Leslie.

"And you might as well be hung for an ox as an ass," said Margo.

"Or your neighbor's wife, if it comes to that," said Larry.

"Now it's a question of whom to invite," said Leslie.

"Theodore, of course," said the family in unison.

"Then there's poor old Creech," said Larry.

"Oh, no, Larry," Mother said, "you know what a disgusting old brute he is."

"Nonsense, Mother, the old boy loves a party."

"And then there's Colonel Ribbindane," said Leslie.

"No!" Larry exclaimed vehemently. "We're not having that quintessence of boredom, even if he is the best shot on the island."

"He's not a bore," said Leslie belligerently. "He's no more boring than your bloody friends."

"None of my friends are capable of spending an entire evening telling you in words of one syllable and a few Neanderthal grunts how he shot a hippo on the Nile in 1904," said Larry coldly.

"It's jolly interesting," said Leslie hotly, "damned sight more interesting than listening to all your friends going on about bloody art."

"Now, now, dears," said Mother peaceably, "there'll be plenty of room for everyone."

I left them to the normal uproar of vituperation that went on over the guest list for any party; as far as I was concerned, as long as Theodore was coming, the party was assured of success, so I could leave the choice of other guests to my family.

Gradually the preparations for the party gathered momentum. Larry succeeded in borrowing Countess Lefraki's enormous grand piano and a tiger-skin rug to place alongside it. The piano was conveyed to us with the utmost tenderness, for it had been the favorite instrument of the late count, on the back of a long flat cart drawn by four horses. Larry, who had been to supervise the removal personally, took off the tarpaulins covering the instrument against the sun, mounted the cart and ran off a quick rendering of "Walking My Baby Back Home" to make sure the piano had not suffered from its journey. It seemed in good shape, if a trifle jangly, and after a prodigious effort we managed to get it into the drawing room. Planted, black and gleaming as an agate, in the corner, the magnificent tiger skin lying in front of it, the head snarling in defiance, it gave the whole room a rich, oriental air.

This was added to by Margo's decorations – great tapestries that she had painted on huge sheets of paper and hung on the walls, pictures of minarets, peacocks, cupolaed palaces and bejeweled elephants, and everywhere vases of ostrich feathers dyed all the colors of the rainbow and bunches of multicolored balloons like crops of strange tropical fruit. The kitchen, of course, was like the interior of Vesuvius; in the flickering ruby light of half a dozen charcoal fires, Mother and her minions scurried to and fro, and the sound of beating and chopping and stirring was so loud it precluded speech, while the aromatic smells that drifted upstairs were so rich and heavy it was like being wrapped in an embroidered cloak of scent.

Over all this Spiro presided, like a scowling brown genie; he seemed to be everywhere, bull-voiced, barrel-bodied, carrying enormous boxes of food and fruit to the kitchen in his hamlike hands, sweating and roaring and cursing as three dining tables were insinuated into the dining room and joined together, appearing with everlasting flowers for Margo and strange spices and other delicacies for Mother. It was during moments like this that you realized Spiro's true worth, for you could ask the impossible of him and he would achieve it. "I'll fixes that," he would say, and fix it he would, whether it was obtaining out-of-season fruit or procuring such a thing as a piano tuner, a species of human being that had been extinct on the island since 1890, as far as everyone knew. It was extremely unlikely, in fact, that any of our parties would have got beyond the planning stage if it had not been for Spiro.

At last everything was ready. The sliding doors between the dining room and drawing room had been pulled back, and the vast room thus formed was a riot of flowers, balloons and paintings, the long tables with their frost-white cloths sparkling with silver, the side tables groaning under the weight of the cold dishes. Sucking pig, brown and polished as a mummy, with an orange in its mouth, lay beside a haunch of wild boar, sticky with wine and honey marinade, thick with pearls of garlic and the round seeds of coriander; a bank of biscuit-brown chickens and young turkeys was interspersed with wild ducks stuffed with wild rice, almonds and sultanas, and woodcock skewered on lengths of bamboo; mounds of saffron rice, yellow as a summer moon, were treasure troves that made one feel like an archaeologist, so thickly were they encrusted with fragile pink strips of octopus, toasted almonds and walnuts, tiny green grapes, carunculated hunks of ginger and pine seeds. The *kefalia* I had brought from the lake were now browned and charcoal-blistered, gleaming in a coating of oil and lemon juice, spattered with jade-green flecks of fennel; they lay in ranks on the huge plates, looking like a flotilla of strange boats tied up in harbor. Interspersed with all this were the plates of small things – crystallized orange and lemon rind, sweet corn, flat thin oat cakes gleaming with diamonds of sea salt, chutney and pickles in a dozen colors and smells and tastes to tantalize and soothe the taste buds. Here was the peak of the culinary art; here a hundred strange roots and seeds had given up their sweet essence; vegetables and fruits had given their rinds and flesh to wash the fowl and the fish in layers of delicately scented gravies and marinades. The stomach twitched at this bank of edible color and smell; you felt you would be eating a magnificent garden, a multicolored tapestry, and you felt that the cells of your lungs would be so filled with layer upon layer of fragrance that you would be drugged and immobile, like a beetle in the heart of a rose. The dogs and I tiptoed several times into the room to look at this succulent display; we would stand until the saliva filled our mouths and then reluctantly go away. We could hardly wait for the party.

Jeejee, whose boat had been delayed, arrived on the morning of his birthday, dressed in a ravishing peacock-blue outfit, his turban immaculate. He was leaning rather heavily on a stick but otherwise showed no signs of his accident, and was as ebullient as ever. To our embarrassment, when showed the preparations we had made for his birthday, he burst into tears.

"To think that I, the son of a humble sweeper, an untouchable, should be treated like this," he sobbed.

"Oh, it's nothing really," said Mother, rather alarmed by his reaction. "We often have little parties."

As our living room looked like a cross between a successful Roman banquet and the Chelsea Flower Show, this gave the impression that we always entertained on a scale that would have been envied by the Tudor court.

"Nonsense, Jeejee," said Larry. "You an untouchable! Your father was a lawyer."

"Vell," said Jeejee, drying his eyes, "I vould have been untouchable if my father had been a different caste. The trouble vith you, Lawrence, is that you have no sense of the dramatic. Think vhat a poem I could have vritten – 'The Untouchable's Banquet.'"

"What's an untouchable?" Margo asked Leslie in a penetrating whisper.

"It's a disease, like leprosy," he said solemnly.

"My God!" said Margo dramatically. "I hope he's sure he hasn't got it. How does he know his father isn't infected?"

"Margo, dear," said Mother quellingly, "go and stir the lentils, will you?"

We had a riotous picnic lunch on the verandah. Jeejee regaled us with stories of his trip to Persia and sang Persian love songs to Margo with such verve that all the dogs howled in unison.

"Oh, you *must* sing one of those tonight," said Margo, delighted, "you must, Jeejee. Everyone's going to do something."

"Vat you mean, Margo dear?" asked Jeejee, mystified.

"We've never done it before. It's a sort of cabaret. Everyone's going to do something," Margo explained. "Lena's going to do a bit of opera – something out of the *Rosy Cavalier*. Theodore and Kralefsky are going to do a trick by Houdini . . . you know, everyone's going to do something . . . so you must sing in Persian."

"Vy couldn't I do something more in keeping with

Mother India?" said Jeejee, struck by the thought. "I could levitate."

"No," said Mother, interrupting firmly, "I want this party to be a success. No levitation."

"Why don't you be something typically Indian?" said Margo. "I know, be a snake charmer!"

"Yes," said Larry, "the humble, typical, untouchable Indian snake charmer."

"My God! Vat a vonderful idea!" said Jeejee, his eyes shining. "I vill do so."

Anxious to be of service, I said I could lend him a basket of small and harmless slowworms for his act, and he was delighted with the idea that he would actually have some real snakes to charm. Then we all went to siesta and to prepare ourselves for the great evening.

The sky was striped green, pink and smoke-grey, and the first owls had started to chime in the dark olives when the guests began to arrive. Among the first was Lena, clasping a huge book of operatic music under her arm and wearing a flamboyant evening dress of orange silk in spite of the fact that she knew the party was informal.

"My dears," she said thrillingly, her black eyes flashing, "I'm in great voice tonight. I feel I shall do justice to the master. No, no, not ouzo, it might afflict my vocal cord. I will have a tiny champagne and brandy. Yes, I can feel my throat vibrate, you know? – like a harp."

"How nice," said Mother insincerely. "I'm sure we shall all enjoy it."

"She's got a lovely voice, Mother," said Margo. "It's a mezzotint."

"Soprano," said Lena coldly.

Theodore and Kralefsky arrived together, carrying a coil of ropes and chains and several padlocks.

"I hope," said Theodore, rocking up and down on his toes, "I hope our . . . er . . . little . . . you know . . . our little illusion will be successful. We have, of course, never done it before."

"*I* have done it before," said Kralefsky with dignity. "It was Houdini himself who showed me. He even went so far as to compliment me on my dexterity. 'Richard,' he said – for we were on intimate terms, you understand, 'Richard, I've never seen anyone except myself so nimble-fingered.'"

"Really?" said Mother. "Well, I'm sure it will be a great success."

Captain Creech arrived wearing a battered top hat, his face strawberry-red, his thistledown hair looking as though the slightest breeze would blow it from his head and chin. He staggered more than usual, and his broken jaw looked more lopsided than normal; it was obvious that he had been priming himself well prior to his arrival. Mother stiffened and gave a forced smile as he lurched through the front door.

"My! You look really sumptuous tonight," said the captain, leering at Mother and rubbing his hands and swaying gently. "You've put on some weight lately, haven't you?"

"I don't think so," said Mother primly.

The captain eyed her up and down critically.

"Well, you seem to have a better handful in your bustle than you used to have," he said.

"I would be glad if you would refrain from making personal remarks, Captain," said Mother coldly.

The captain was unabashed.

"It doesn't worry *me*," he confided. "I like a woman with a bit of something you can get your hands on. A thin woman's no good in bed – like riding a horse with no saddle."

"I have no interest in your preference either in or out of bed," said Mother with asperity.

"Well, there are plenty of other places," said Captain Creech accommodatingly. "I knew a wench once who could do a marvelous job on a camel. Bedouin Bertha, they called her."

"I would be glad if you'd keep your reminiscences to yourself, Captain," said Mother crushingly, looking round desperately for Larry.

"I thought you'd be interested," said Captain Creech, surprised. "Back of a camel's a difficult thing – real specialist job."

"I'm not interested in how specialized your female acquaintances are," said Mother. "Now, if you'll excuse me, I have to go and attend to the food."

More and more carriages clopped up to the front door, and more and more cars disgorged guests. The room filled up with the strange selection of people the family had invited. In one corner Kralefsky, like an earnest humpbacked gnome, was telling Lena about his experiences with Houdini.

"'Harry,' I said to him – for we were intimate friends, you understand, 'Harry, show me what secrets you like, they are safe with me. My lips are sealed.'"

Kralefsky took a sip of his wine and pursed his lips to show how they were sealed.

"Really?" said Lena, with total lack of interest. "Vell, of course it's different in the singing vorld. Ve singers pass on our secrets. I remember Krasia Toupti saying to me, 'Lena your voice is so beautiful I cry vhen I hear it; I have taught you all know. Go, carry the torches of our genius to the vorld.'"

"I didn't mean to imply that Harry Houdini was secretive," said Kralefsky stiffly. "He was the most generous of men. Why, he even showed me how to saw a woman in half."

"My dear, how curious it must feel to be cut in half," said Lena musingly. "Think of it, your bottom half could be having an affair in one room while your other half was entertaining an archbishop. How droll."

"It's only an illusion," said Kralefsky, going pink.

"So is life," said Lena soulfully, "so is life, my friend."

The noise of drinking was exhilarating. Champagne corks popped, and the pale, chrysanthemum-colored liquid, whispering gleefully with bubbles, hissed into the glasses; heavy red wine glupped into the goblets, thick and crimson as the blood of some mythical monster, and a swirling wreath of pink bubbles formed on the surface; the frosty white wine tiptoed into the glasses, shrilling, gleaming, now like diamonds, now like topaz; the ouzo lay transparent and innocent as the edge of a mountain pool until the water splashed in and the whole glass curdled like a conjuring trick, coiling and blurring into a summer cloud of moonstone white.

Presently we moved down the room to where the vast array of food awaited us, and the king's butler, fragile as a mantis, superintended the peasant girls in the serving. Spiro, scowling more than usual with concentration, meticulously carved the joints and the birds. Kralefsky had been trapped by the great, grey, walruslike bulk of Colonel Ribbindane, who loomed over him, his giant moustache hanging like a curtain over his mouth, his bulbous blue eyes fixed on Kralefsky in a paralyzing stare.

"The hippopotamus, or river horse, is one of the largest of the quadrupeds to be found in the continent of Africa. . . ." he droned, as though lecturing a class.

"Yes, yes . . . fantastic beast. Truly one of nature's wonders," said Kralefsky, looking round desperately for escape.

"When you shoot a hippopotamus, or river horse," droned Colonel Ribbindane, oblivious to interruption, "as I have had the good fortune to do, you aim between the eyes and the ears, thus ensuring that the bullet penetrates the brain."

"Yes, yes," Kralefsky agreed, hypnotized by the colonel's protuberant blue eyes.

"Bang!" said the colonel so suddenly and loudly that Kralefsky nearly dropped his plate. "You hit him between the eyes. Splash! Crunch! Straight into the brain, d'you see?"

"Yes, yes," said Kralefsky, swallowing and going white.

"Splosh!" said the colonel, driving the point home. "Blow his brains out in a fountain."

Kralefsky closed his eyes in horror and put his half-eaten plate of suckling pig down.

"He sinks then," the colonel went on, "sinks right down to the bottom of the river . . . glug, glug, glug. Then you wait twenty-four hours – d'you know why?"

"No . . . I . . . uh . . . " said Kralefsky, swallowing frantically.

"Flatulence," said the colonel with satisfaction. "All the semidigested food in its belly, d'you see? It rots and produces gas. Up puffs the old belly like a balloon and up she pops."

"H-How interesting," said Kralefsky faintly. "I think, if you will just excuse me. . . ."

"Funny things, stomach contents," mused the colonel, ignoring Kralefsky's attempts at escape. "Belly is swollen up to twice its natural size; when you cut it open, whoosh! like slicing up a zeppelin full of sewage, d'you see?"

Kralefsky put his handkerchief over his mouth and gazed round in an anguished manner.

"Different with the elephant, the *largest* land quadruped in Africa," the colonel droned on, filling his mouth with crisp suckling pig. "D'you know, the Pygmies cut it open and crawl into the belly and eat the liver all raw and bloody . . . still quivering sometimes. Funny little chaps, Pygmies . . . wogs, of course. . . ."

Kralefsky, now a delicate shade of yellow-green, escaped to the verandah, where he stood in the moonlight taking deep breaths.

The suckling pig had vanished, the bones gleamed white in the joints of lamb and boar, and the rib cages and breastbones of the chickens and turkeys and ducks lay like the wreckage of upturned boats. Jeejee, having sampled a little of everything, at Mother's insistence, and having declared it infinitely superior to anything he had ever eaten before, was vying with Theodore to see how many Taj Mahal Tidbits they could consume.

"Delicious," said Jeejee indistinctly, his mouth full, "simply delicious, my dear Mrs. Durrell. You are the apotheosis of culinary genius."

"Yes indeed," said Theodore, popping another Taj Mahal Tidbit into his mouth and scrunching it up, "they're really excellent. They make something similar in Macedonia . . . er . . . um . . . but with goat's milk."

"Jeejee, did you really break your leg levitizing, or whatever it's called?" asked Margo.

"No," said Jeejee sorrowfully. "I vouldn't mind if I had. It vould have been in a good cause. No, the damned stupid hotel vere I stayed had French vindows in the bedrooms but they couldn't afford a balcony."

"Sounds like a Corfu hotel," said Leslie.

"So one evening I vas overcome with forgetfulness and I stepped out onto the balcony to do some deep breathing; and of course there vas no balcony."

"You might have been killed," said Mother. "Have another tidbit."

"Vat is death?" asked Jeejee oratorically. "A mere sloughing of the skin, a metamorphosis. I vent into a deep trance in Persia, and my friend got incontrovertible proof that in a previous life I vas Genghis Khan."

"You mean the film star?" asked Margo, wide-eyed.

"No, dear Margo, the great varrior," said Jeejee.

"You mean you could remember being him?" asked Leslie, interested.

"Alas, no. I vas in a trance," said Jeejee sadly. "One is not allowed to remember one's previous lives."

"You . . . khan have your cake and eat it," explained Theodore, delighted at having found an opportunity for a pun.

"I wish everybody would hurry up and finish eating," said Margo. "Then we can get on with the acts."

"To hurry such a meal vould be an insult," said Jeejee. "There is time; the whole night stretches before us. Besides, Gerry and I have to go and organize my supporting cast of reptiles."

It took quite some time before the cabaret was ready, for everyone was full of wine and good food and refused to be hurried; eventually, however, Margo got the cast assembled. She had tried to get Larry to be master of ceremonies, but he had refused. He said that if she wanted him to be part of the cabaret he was not going to be master of ceremonies as well, so in desperation she had had to step into the breach herself. Blushing slightly, she took up her place on the tiger skin by the piano and called for silence.

"Ladies and gentlemen," she said, "tonight, for your entertainment, we have a cabaret of the best talent on the island, and I'm sure that you will all enjoy the talents of these talented talents."

She paused, blushing, while Kralefsky gallantly led the applause.

"First I would like to introduce Constantino Megalotopolopopoulos," she continued, "who is going to act as accompanist."

A tiny, fat little Greek, looking like a swarthy ladybird, trotted into the room, bowed and sat down at the piano. This had been one of Spiro's achievements, for Mr. Megalotopolopopoulos, a draper's assistant, could not only play the piano but read music as well.

"And now," said Margo, "it is with great pleasure that I present to you that very talented artiste Lena Mavrokondas and Constantino Megalotopolopopoulos on the piano. Lena will sing that great aria from *Rosy Cavalier*, 'The Presentation of the Rose.'"

Lena, glowing like a tiger lily, swept to the piano, bowed to Constantino, placed her hands carefully over her midriff as though warding off a blow, and commenced to sing.

"Beautiful, beautiful," said Kralefsky as she finished and bowed to our applause. "What virtuosity."

"Yes," said Larry, "it used to be known as the thrice-vee method at Covent Garden."

"Three-vee?" asked Kralefsky, much interested. "What's that?"

"Vim, vibrato and volume," said Larry.

"Tell them I will sing encore," whispered Lena to Margo after whispered consultation with Constantino Megalotopolopopoulos.

"Oh, yes. How nice," said Margo, flustered and unprepared for this largesse. "Ladies and gentlemen, Lena will now sing another song, called 'The Encore.'"

Lena gave Margo a withering look and swept into her next song with such vigor and so many gestures that even Creech was impressed.

"By George, she's a good-looking wench, that!" he exclaimed, his eyes watering with enthusiasm.

"Yes, a true artiste," agreed Kralefsky.

"What chest expansion," said Creech admiringly. "Bows like a battleship."

Lena finished on a zitherlike note and bowed to the applause, which was loud but nicely judged in warmth and length to discourage another encore.

"Thank you, Lena, that was wonderful. Just like the real thing," said Margo, beaming. "And now, ladies and gentlemen, I present the famous escape artists, Krafty Kralefsky and his partner, Slithery Stephanides."

"Dear God," said Larry, "who thought of those names?"

"Need you ask?" said Leslie. "Theodore. Kralefsky wanted to call the act the Mysterious Escapologist Illusionists, but Margo couldn't guarantee to say it properly."

"One must be thankful for small mercies," said Larry.

Theodore and Kralefsky clanked onto the floor near the piano carrying their load of ropes, chains and padlocks.

"Ladies and gentlemen," said Kralefsky, "tonight we will show you tricks that will baffle you, tricks so mysterious that you will be agog to know how they are done."

He paused to frown at Theodore, who had dropped a chain on the floor by mistake.

"For my first trick, I will ask my assistant not only to bind me securely with rope but chain as well."

We clapped dutifully and watched, delighted, while Theodore wound yards and yards of rope and chain around Kralefsky. Occasional whispered altercations drifted to the audience.

"I've . . . er . . . you know . . . um . . . forgotten precisely the knot . . . Um . . . yes . . . you mean the padlock *first?* Ah yes, I have it . . . hm . . . er . . . just a second."

At length Theodore turned to the audience and said, "I must apologize for . . . er . . . you know . . . er . . . taking so long, but, unfortunately, we didn't have time to . . . er . . . practice, that is to say – "

"Get on with it!" hissed Krafty Kralefsky.

Theodore had wound so many lengths of rope and chain around Kralefsky that he looked as though he had stepped straight out of Tutankhamun's tomb.

"And now," said Theodore, with a gesture at the immobile Kralefsky, "would anyone like to . . . er . . . you know . . . examine the knots?"

Colonel Ribbindane lumbered forward.

"Er . . . um . . . " said Theodore, startled, not having expected his offer to be taken up, "I'm afraid I must ask you to . . . um . . . that is to say . . . if you don't actually *pull* on the knots . . . er . . . um . . . "

Colonel Ribbindane made an inspection of the knots that was so minute one would have thought he was chief warder in a prison. At length, and with obvious reluctance, he pronounced the knots good. Theodore looked relieved as he stepped forward and gestured at Kralefsky again.

"And now, my assistant, that is to say, my *partner*, will show you how . . . easy it is to . . . er . . . you know . . . um . . . rid yourself of . . . er . . . um . . . several yards . . . feet, I should say . . . though, of course, being in Greece, perhaps one should say meters . . . er . . . um . . . several meters of . . . er . . . rope and chains."

He stepped back and we all focused our attention on Kralefsky.

"Screen!" he hissed at Theodore.

"Ah! Hm . . . yes," said Theodore and laboriously moved a screen in front of Kralefsky.

There was a long and ominous pause, during which

we could hear panting and the clanking of chains from behind the screen.

"Oh dear," said Margo, "I do hope he can do it."

"Shouldn't think so," said Leslie. "All those padlocks look rusty."

But at the moment, to our astonishment, Theodore whipped away the screen and revealed Kralefsky, slightly purple of face and disheveled, standing free in a pool of ropes and chains.

The applause was genuine and surprised, and Kralefsky flushed and basked in the adulation of his audience.

"My next trick, a difficult and dangerous one, will take some time," he said portentously. "I shall be roped and chained by my assistant, and the knots can be examined by – ha! ha! – the sceptics among you, and then I shall be cast into an airtight box. In due course you will see me emerge miraculously, but it takes some time for me to achieve this . . . er . . . miracle. The next act will kindly entertain you."

Spiro and Megalotopolopopoulos appeared, dragging a large and extremely heavy olive-wood chest of the sort that used to be used for keeping linen in. It was ideal for the purpose, for when Kralefsky had been roped and chained and the knots had been examined minutely by a suspicious Colonel Ribbindane, Kralefsky was lifted into it by Theodore and Spiro and he slid into the interior as neatly as a snail into its shell. Theodore, with a flourish, slammed the lid shut and locked it.

"Now, when my assis . . . er . . . my . . . er . . . um . . . partner, that is to say . . . signals me, I will release him," he said. "On with the show!"

"I don't like it," said Mother. "I hope Mr. Kralefsky knows what he's doing."

"I shouldn't think so," said Leslie gloomily.

"It's too much like . . . well . . . premature burial," said Mother.

"Perhaps when we open it he'll have changed into Edgar Allan Poe," said Larry hopefully.

"It's perfectly all right, Mrs. Durrell," said Theodore. "I can communicate with him by a series of knocks . . . um . . . a sort of Morse code."

"And now," said Margo, "while we are waiting for Krafty Kralefsky to escape, we have that incredible snake charmer from the East, Prince Jeejeebuoy."

Megalotopolopopoulos played a series of thrilling chords and Jeejee trotted into the room. He had removed his finery and was clad simply in a turban and loincloth. As he could not find a suitable snake-charming pipe, he was carrying a violin which he had got Spiro to borrow from a man in the village; in his other hand he held his basket containing his act. He had rejected with scorn my slowworms when he had seen them as being far too small to aid in the cultivation of the image of Mother India. He had insisted instead on borrowing one of my water snakes, an elderly specimen some two and a half feet long and of an extremely misanthropic disposition. As he bowed to the audience, the top fell off the basket and the snake, looking very disgruntled, fell out onto the floor. Everyone panicked except Jeejee, who squatted down cross-legged near the coiled snake, tucked the violin under his chin and started to play. Gradually the panic subsided, and we all watched entranced as Jeejee swayed to and fro extracting the most agonizing noises from the fiddle, watched by the alert and irritated snake. Just at that moment came a loud knock from the box in which Kralefsky was incarcerated.

"Aha!" said Theodore. "The signal!"

He went to the box and bent over, his beard bristling

as he tapped on it like a woodpecker. Everyone's attention was on him, including Jeejee's, and at that moment the water snake struck. Fortunately, Jeejee moved so that the water snake only got a firm hold on his loincloth; however, it hung on grimly and pugnaciously.

"Ow! My God!" screamed the incredible snake charmer from the East. "Hey, Gerry, quick! Quick, it's biting me in the crutch."

It was some seconds before I could persuade him to stand still so that I could disentangle the snake from his loincloth. During this time Theodore was having a prolonged Morse code conversation with Kralefsky in the box.

"I do not think I can do any more," said Jeejee, shakily accepting a large brandy from Mother. "It tried to bite me below the belt!"

"He will apparently be a minute or two yet," announced Theodore. "He's had a little trouble . . . er . . . difficulty, that is, with the padlocks. At least, that's what I understood him to say."

"I'll put the next act on," said Margo.

"Think," said Jeejee faintly, "it might well have been a cobra."

"No, no," said Theodore, "cobras are not found here in Corfu."

"And now," said Margo, "we have Captain Creech, who will give us some old-time songs, and I'm sure you'll want to join in with him. Captain Creech."

The captain, his top hat tilted at a rakish angle, strutted across to the piano and did a little bowlegged to-and-fro shuffle, twirling the cane he had procured.

"Old sea chantey," he bellowed, putting his top hat on the end of his cane and twirling it round dexterously, "old sea chantey. You all join in the chorus."

He did a short dance, still twirling his hat, and came

in on the beat of the song which Megalotopolopopoulos was thumping out.

> *"Oh, Paddy was an Irishman,*
> *He came from Donegal,*
> *And all the girls they loved him well,*
> *Though he only had one ball,*
> *For the Irish girls are girls of sense,*
> *And they didn't mind at all,*
> *For, as Paddy pointed out to them,*
> *'Twas better than none at all.*
> *Oh, folderol and folderay,*
> *A sailor's life is grim,*
> *So you're only too delighted,*
> *If you get a bit excited,*
> *Whether it's with her or him."*

"Really, Larry!" said Mother, outraged. "Is this your idea of entertainment?"

"Why pick on me?" said Larry, astounded. "It's nothing to do with me."

"You invited him, disgusting old man. He's your friend."

"I can't be responsible for what he *sings*, can I?" asked Larry irritably.

"You must put a stop to it," said Mother. "Horrible old man."

"He certainly twirls his hat round very well," said Theodore enviously. "I wonder how ... he ... er ... does it?"

"I'm not interested in his hat. It's his songs," said Mother.

"It's a perfectly good music hall ditty," said Larry. "I don't know what you're going on about."

"It's not the sort of music hall ditty *I'm* used to," said Mother.

"Oh, Blodwyn was a Welsh girl,
She came from Cardiff city,
And all the boys they loved her well,
Though she only had one titty,"

caroled the captain, getting into his stride.

"Repulsive old fool!" said Mother.

"For the Welsh boys there
Are boys of sense,
And didn't they all agree,
One titty is better than two sometimes,
For it leaves you one hand free.
Oh, folderol and folderay,
A sailor's life is grim . . . "

"Even if you don't consider me, you might consider Gerry," said Mother.

"What d'you want me to do? Write the verses down for him?" asked Larry.

"D'you . . . you know . . . hear a sort of *tapping*?" asked Theodore.

"Don't be ridiculous, Larry, you know perfectly well what I mean."

"I wondered if he might be ready . . . um . . . the trouble is, I can't quite remember the signal," Theodore confessed.

"I don't know why you always have to pick on me," said Larry, "just because *you're* narrow-minded."

"I'm as broad-minded as anybody," said Mother indignantly, "in fact, sometimes I think I'm too broad."

"I *think* it was two slow and three quick," mused Theodore, "but I may be mistaken."

"Oh, Gertrude was an English lass,
She came from Stoke-an-Trent,
But when she loved a nice young lad,
She always left him bent."

"Listen to that!" said Mother. "It's beyond a joke. Larry, you must stop him."

"*You're* objecting, *you* stop him," said Larry.

"But the boys of Stoke,
They loved a poke,
And suffered in the bed,
For they said that Gert
Was a real prime skirt,
But she had a left-hand thread."

"Really, Larry, you carry things too far. It's not funny."

"Well, he's been through Ireland, Wales and England," Larry pointed out. "He's only got Scotland to go, unless he branches out into Europe."

"You must stop him doing that!" said Mother, aghast at the thought.

"I think, you know, perhaps I ought to just open the box and have a look," said Theodore thoughtfully. "You know, just as a *precaution*."

"I wish you'd stop carrying on like a female Bowdler," said Larry. "It's all good clean fun."

"Well, it's not *my* idea of good clean fun," said Mother, "and I want it stopped."

"Oh, Angus was a Scottish lad,
He came from Aberdeen . . . "

"There you are, he's got to Scotland now," said Larry.

"Er . . . I'll try not to disturb the captain," said Theodore, "but I thought perhaps just to take a quick glance – "

"I don't care whether he's got to John o'Groats," said Mother, "it's got to stop."

Theodore had tiptoed over to the box and was now feeling in his pockets worriedly. Leslie joined him and they discussed the problem of the entombed Kralefsky. I saw Leslie trying ineffectually to raise the lid when it became obvious that Theodore had lost the key.

The captain sang on unabated.

"Oh, Fritz, he was a German lad,
He came from old Berlin . . . "

"There!" said Mother. "He's started on the Continent! Larry, you must stop him!"

"I wish you'd stop carrying on like the lord chamberlain," said Larry, annoyed. "It's Margo's cabaret; tell *her* to stop him."

"It's a mercy that most of the guests don't speak good enough English to understand," said Mother, "though what the others must think – "

"Folderol and folderay,
A sailor's life is grim . . . "

"I'd make life grim for him if I could," said Mother, "depraved old fool!"

Leslie and Theodore had now been joined by Spiro, carrying a large crowbar; together they set about the task of trying to open the lid.

"Oh, Françoise was a French girl,
She came from the town of Brest,
And, oh, she lived up to its name,
And gave the boys no rest."

"I do try to be broad-minded," said Mother, "but there *are* limits."

"Tell me, my dears," asked Lena, who had been listening to the captain with care, "what is left-hand thread?"

"It's a . . . it's a . . . it's a sort of English joke," said Mother desperately, "like a pun, you know?"

"Yes," explained Larry, "like you say a girl's got a pun in the oven."

"Larry, that's quite enough," said Mother quellingly. "The captain's bad enough without *you* starting."

"Mother," said Margo, "I think Kralefsky's suffocating."

"I do not understand this pun in oven," said Lena. "Explain me."

"Take no notice, Lena. It's only Larry's joke."

"If he's suffocating, ought I to go and stop the captain's song?" asked Margo.

"An excellent idea! Go and stop him at once," said Mother.

There were loud groaning noises as Leslie and Spiro struggled with the heavy lid of the chest. Margo rushed up to the captain.

"Captain, captain, please stop," she said. "Mr. Kralefsky's . . . Well, we're rather worried about him."

"Stop?" said the captain, startled. "Stop? But I've only just begun."

"Yes, well, there are more urgent things than your song," said Mother frigidly. "Mr. Kralefsky's stuck in his box."

"But it's one of the best songs I know," said the captain aggrievedly. "It's the longest, too – one hundred and forty countries it deals with – Chile, Australia, the Far East, the lot. A hundred and forty verses!"

I saw Mother flinch at the thought of the captain singing the other hundred and thirty-four verses.

"Yes, well, some other time perhaps," she said untruthfully, "but this is an emergency."

With a splintering noise like a giant tree being felled, the lid of the chest was finally wrenched open, and inside lay Kralefsky, still swathed in ropes and chains, his face an interesting shade of blue, his hazel eyes wide and terrified.

"Aha, I see we're a bit . . . er . . . you know . . . premature," said Theodore. "He hasn't succeeded in untying himself."

"Air! Air!" croaked Kralefsky. "Give me air!"

"Interesting," said Colonel Ribbindane. "Saw a Pygmy like that once in the Congo . . . been trapped in an elephant's stomach. The elephant is the largest African quadruped – "

"Do get him out," said Mother agitatedly. "Get some brandy."

"Fan him! Blow on him!" shrilled Margo, and burst into tears. "He's dying, he's dying, and he never finished his trick."

"Air . . . air," moaned Kralefsky as they lifted him out of the box.

In his shroud of ropes and chains, his face leaden, his eyes closed, he certainly looked a macabre sight.

"I think perhaps, you know, the ropes and chains are a little constricting," said Theodore judiciously, becoming the medical man.

"Well, *you* put them on him; now get them off him,"

said Larry. "Come on, Theodore, you've got the key for the padlocks."

"I seem, rather unfortunately, to have mislaid it," Theodore confessed.

"Dear God!" said Leslie. "I knew they shouldn't be allowed to do this. Damned silly. Spiro, can you get a hacksaw?"

They laid Kralefsky on the sofa and supported his head on the cushions. He opened his eyes and gasped at us helplessly. Colonel Ribbindane bent and stared into Kralefsky's face.

"This Pygmy I was telling you about," he said, "his eyeballs filled up with blood."

"Really?" said Theodore, much interested. "I believe you get the same when someone is . . . er . . . you know . . . garroted. A rupturing of the blood vessels in the eyeballs sometimes bursts them."

Kralefsky gave a small, despairing squeak like a field mouse.

"Now, if he had taken a course in Fakyo," said Jeejee,"he vould have been able to hold his breath for hours, perhaps even days, possibly even months or *years*, vith practice."

"Would that prevent his eyeballs' filling with blood?" asked Ribbindane.

"I don't know," said Jeejee honestly. "I'm sure it vould prevent them filling with blood; they'd probably just go pink."

"Are my eyes full of blood?" asked Kralefsky agitatedly.

"No, no, of course they're not," said Mother soothingly. "I do wish you all would stop talking about blood and worrying poor Mr. Kralefsky."

"Yes, take his mind off it," said Captain Creech. "Shall I finish my song?"

"No," said Mother firmly, "no more songs. Why don't you get Mr. Maga . . . whatever his name is to play something soothing and all have a nice dance while we unwrap Mr. Kralefsky?"

"That's an idea, my lovely wench," said Captain Creech to Mother. "Waltz with me! One of the quickest ways of getting intimate, waltzing."

"No," said Mother coldly. "I'm much too busy to get intimate with anyone, thank you very much."

"You, then," said the captain to Lena, "you'll give me a cuddle round the floor, huh?"

"Vell, I must confess it, I like the valtz," said Lena, puffing out her chest to the captain's obvious delight.

Megalotopolopopoulos swung himself into a spirited rendering of "The Blue Danube" and the captain whisked Lena off across the room.

"The trick *would* have worked perfectly, only Dr. Stephanides should have only *pretended* to lock the padlocks," Mr. Kralefsky was explaining, while a scowling Spiro hacksawed away at the locks and chains.

"Of course," said Mother, "we quite understand."

"I was never . . . er . . . you know . . . very good at conjuring," said Theodore contritely.

"I could feel the air running out and I could hear my heartbeats getting louder and louder. It was horrible, quite horrible," said Kralefsky, closing his eyes with a shudder that made all his chains jangle. "I began to think I'd never get out."

"And you missed the rest of the cabaret too," said Margo sympathetically.

"Yes, by God!" exclaimed Jeejee. "You didn't see my snake charming. Damned great snake bit me in the loincloth, and me an unmarried man!"

"And then the blood started pounding in my ears,"

said Kralefsky, hoping to retain the focus of attention. "Everything went black."

"But . . . er . . . you know . . . it *was* dark in there," Theodore observed.

"Don't be so literal, Theo," said Larry. "One can't embroider a story properly with you damned scientists around."

"I'm not embroidering," said Kralefsky with dignity, as the last padlock fell away and he could sit up. "Thank you, Spiro. No, I assure you, everything went as black as . . . as black as. . . ."

"A nigger's bottom?" offered Jeejee helpfully.

"Jeejee, dear, don't use that word," said Mother, shocked. "It's not polite."

"What? Bottom?" asked Jeejee, mystified.

"No, no," said Mother, "that other word."

"What? Nigger?" he asked. "But vhat's vrong with it? I'm the only nigger here and I don't object."

"Spoken like a white man," said Colonel Ribbindane admiringly.

"Well, I object," said Mother firmly. "I won't have you calling yourself a nigger. To me, you're just as, just as. . . ."

"White as driven snow?" suggested Larry.

"You know perfectly well what I mean, Larry," said Mother crossly.

"Well," said Kralefsky, "there was I with the blood pounding in my ears. . . ."

"Oooh," squeaked Margo suddenly, "just look what Captain Creech has done to Lena's lovely dress."

We turned to look at that section of the room where several couples were gyrating merrily to the waltz, none with greater enthusiasm than the captain and Lena. Unfortunately, unbeknownst to either of them, the captain at some point must have trodden on the deep layer of frills that decorated the edge of Lena's gown and wrenched them away; now they were waltzing away oblivious to the fact that the captain had both feet inside Lena's dress.

"Good heavens! Disgusting old man!" said Mother.

"He was right about the waltz being intimate," said Larry. "Another couple of whirls and they'll be wearing the same dress."

"D'you think I ought to tell Lena?" asked Margo.

"I shouldn't," said Larry. "It's probably the nearest she's been to a man in years."

"Larry, that's quite unnecessary," said Mother.

Just at that moment, with a flourish, Megalotopolopopoulos brought the waltz to an end and Lena and the captain spun round and round like a top and then stopped. Before Margo could say anything, the captain stepped backwards to bow and fell flat on his back, ripping a large section of Lena's skirt away. There was a moment's terrible silence while every eye in the room was riveted, fascinated, on Lena, who stood there frozen. The captain broke the spell, speaking from his recumbent position on the floor.

"My, that's a fine pair of knickers you're wearing," he observed jovially.

Lena uttered what can only be described as a Greek screech, a sound that has all the bloodcurdling qualities of a scythe blade scraped across a hidden rock; a sound, part lamentation, part indignation, with a rich, murderous overtone, a noise wrenched up, as it were, from the very bowels of the vocal cords. Galli-Curci would have been proud of her. It was, strangely enough, Margo who leapt into the breach and circumnavigated what could have been a diplomatic crisis, though her method of doing so was perhaps a trifle flamboyant. She simply snatched a cloth from a side table, rushed to Lena and swathed her in it. This gesture in itself would have been all right except that she chose a cloth on which there were reposing numerous dishes of food and a large twenty-four-branch candelabrum. However, the crash of breaking china and the hissing of candles falling into chutneys and sauces successfully distracted the guests from Lena, and under cover of the pandemonium she was rushed upstairs by Margo.

"I hope you're satisfied now!" said Mother accusingly to Larry.

"Me? What have I done?" he inquired.

"That man," said Mother. "You invited him. Now look what he's done."

"Given Lena the thrill of her life," said Larry. "No man ever tried to tear her skirt off before."

"It's not funny, Larry," said Mother severely, "and if we have any more parties, I will not have that . . . that . . . licentious old libertine."

"Never mind, Mrs. Durrell, it's a lovely party," said Jeejee.

"Well, as long as you're enjoying it, I don't mind," said Mother, mollified.

"If I have a hundred reincarnations, I'm sure I shall never have another birthday party like this."

"That's very sweet of you, Jeejee," said Mother.

"There's only one thing," said Jeejee soulfully. "I hesitate to mention it . . . but. . . ."

"What?" asked Mother. "What was wrong?"

"Not wrong," said Jeejee sighing, "just *lacking*."

"*Lacking?*" asked Mother, alarmed. "What was lacking?"

"Elephants," said Jeejee earnestly, "the largest quadrupeds in India."

AFTERWORD

I HOPE YOU HAVE enjoyed this book, and on the off chance that I've caught you in a weak and impressionable moment, I would like to talk to you about my other consuming passion – breeding endangered animals. If you have read any of my books about wildlife conservation, you will know exactly what it's about. If you haven't, then let me, very briefly, tell you about a number of people who contribute to the Wildlife Preservation Trust International in an attempt to save from extinction some of our rarest animal species in my breeding sanctuary in the English Channel island of Jersey.

You see, animals have given me, personally, so much pleasure over the years that I formed this trust to try to repay this debt in some small measure by using my own experience and that of my staff to breed the world's most critically endangered species as an aid to their conservation in the wild.

We believe that we have no right to stand idly by watching unique life systems being extinguished by our own greed and insensitivity without making one last, desperate effort to preserve viable breeding groups until their natural habitats have been saved and secured and conditions are favorable for their reintroduction.

So, on behalf of the most charming and bizarre, colorful and exotic, fascinating, resourceful, magnificent, dignified, funny and beguiling minority in this world (who cannot read, write, vote or invent nerve gas), I invite you to join us.

Editor's Note:

The Wildlife Preservation Trust International, founded by Gerald Durrell, is now the Durrell Wildlife Conservation Trust. See *www.durrell.org* for more information.

A NOTE ON THE TYPE

FAUNA AND FAMILY *has been set in Fairfield, a type designed by the Czech-born American book designer, typographer, and wood engraver, Rudolph Ruzicka. Best known as an engraver and illustrator, Ruzicka began his study of art at Hull House in Chicago, moving on to study at the Chicago Art Institute and the Art Students League in New York. Ruzicka's early years as an engraver at the American Bank Note Company gave him the opportunity to polish the meticulous sense of craft that would lead to an ever-increasing stream of commissions and, in 1935, the AIGA Medal. Among Ruzicka's most fruitful professional relationships was his long involvement with D. B. Updike and the Merrymount Press, for whom Ruzicka designed a number of books and, famously, engraved the press's annual Christmas card. ❃ A man of formidable typographic knowledge, Ruzicka designed Fairfield during his tenure as director of typographic design at Mergenthaler Linotype. Strongly reminiscent of Walbaum, the type is based in part on the so-called modern types of the nineteenth century, but has little of those earlier types' stiffness. Most suitable when a type of considerable style is called for, Fairfield shows itself to best advantage on clean, simple pages with only a touch of ornamentation.*

DESIGN BY CARL W. SCARBROUGH

COMPOSITION BY CARRIE DIERINGER TAFFEL